The World Without Us

THE
WORLD
WITHOUT
US

ALAN WEISMAN

First published in Great Britain in 2007 by
Virgin Books Ltd
Thames Wharf Studios
Rainville Road
London
W6 9HA

A catalogue record for this book is available from
the British Library

ISBN 978 1 905264 03 2 (HB)
ISBN 978 0 7535 1239 5 (PB)

The paper used in this book is a natural, recyclable product made
from wood grown in sustainable forests. The manufacturing
process conforms to the regulations of the country of origin

Printed and bound in Great Britain by
Mackays of Chatham plc, Chatham, Kent

In memory of
Sonia Marguerite

with lasting love
from a world without you

CONTENTS

PART III

PART IV

Das Firmament blaut ewig, und die Erde
Wird lange fest steh'n und aufblüh'n im Lenz.
Du aber, Mensch, wie lange lebst denn du?

The firmament is blue forever, and the Earth
Will long stand firm and bloom in spring.
But, man, how long will you live?

—Li-Tai-Po/Hans Bethge/Gustav Mahler
   The Chinese Flute:
   Drinking Song of the Sorrow of the Earth
   Das Lied von der Erde

The World Without Us

A Monkey Koan

O NE JUNE MORNING in 2004, Ana María Santi sat against a post beneath a large palm-thatched canopy, frowning as she watched a gathering of her people in Mazáraka, their hamlet on the Río Conambu, an Ecuadoran tributary of the upper Amazon. Except for Ana María's hair, still thick and black after seven decades, everything about her recalled a dried legume pod. Her gray eyes resembled two pale fish trapped in the dark eddies of her face. In a patois of Quichua and a nearly vanished language, Zápara, she scolded her nieces and granddaughters. An hour past dawn, they and everyone in the village except Ana María were already drunk.

The occasion was a *minga*, the Amazonian equivalent of a barn raising. Forty barefoot Zápara Indians, several in face paint, sat jammed in a circle on log benches. To prime the men for going out to slash and burn the forest to clear a new cassava patch for Ana María's brother, they were drinking *chicha*—gallons of it. Even the children slurped ceramic bowls full of the milky, sour beer brewed from cassava pulp, fermented with the saliva of Zápara women who chew wads of it all day. Two girls with grass braided in their hair passed among the throng, refilling *chicha* bowls and serving dishes of catfish gruel. To the elders and guests, they offered hunks of boiled meat, dark as chocolate. But Ana María Santi, the oldest person present, wasn't having any.

Although the rest of the human race was already hurtling into a new millennium, the Zápara had barely entered the Stone Age. Like the spider monkeys from whom they believe themselves descended, the Zápara essentially still inhabit trees, lashing palm trunks together with *bejuco* vines to support roofs woven of palm fronds. Until cassava arrived, palm hearts were their main vegetable. For protein they netted fish and hunted tapirs, peccaries, wood-quail, and curassows with bamboo darts and blowguns.

They still do, but there is little game left. When Ana María's grandparents were young, she says, the forest easily fed them, even though the Zápara were then one of the largest tribes of the Amazon, with some 200,000 members living in villages along all the neighboring rivers. Then something happened far away, and nothing in their world—or anybody's—was ever the same.

What happened was that Henry Ford figured out how to mass-produce automobiles. The demand for inflatable tubes and tires soon found ambitious Europeans heading up every navigable Amazonian stream, claiming land with rubber trees and seizing laborers to tap them. In Ecuador, they were aided by highland Quichua Indians evangelized earlier by Spanish missionaries and happy to help chain the heathen, lowland Zápara men to trees and work them until they fell. Zápara women and girls, taken as breeders or sex slaves, were raped to death.

By the 1920s, rubber plantations in Southeast Asia had undermined the market for wild South American latex. The few hundred Zápara who had managed to hide during the rubber genocide stayed hidden. Some posed as Quichua, living among the enemies who now occupied their lands. Others escaped into Peru. Ecuador's Zápara were officially considered extinct. Then, in 1999, after Peru and Ecuador resolved a long border dispute, a Peruvian Zápara shaman was found walking in the Ecuadoran jungle. He had come, he said, to finally meet his relatives.

The rediscovered Ecuadoran Zápara became an anthropological cause célèbre. The government recognized their territorial rights, albeit to only a shred of their ancestral land, and UNESCO bestowed a grant to revive their culture and save their language. By then, only four members of the tribe still spoke it, Ana María Santi among them. The forest they once knew was mostly gone: from the occupying Quichua they had learned to fell trees with steel machetes and burn the stumps to plant cassava. After a single harvest, each plot had to be fallowed for years; in every direction, the towering forest canopy had been replaced by spindly, second-growth shoots of laurel, magnolia, and *copa* palm. Cassava was now their mainstay, consumed all day in the form of *chicha*. The Zápara had survived into the 21st century, but they had entered it tipsy, and stayed that way.

They still hunted, but men now walked for days without finding tapirs or even quail. They had resorted to shooting spider monkeys, whose flesh was formerly taboo. Again, Ana María pushed away the bowl proffered by her granddaughters, which contained chocolate-colored meat with a tiny, thumbless paw jutting over its side. She raised her knotted chin toward the rejected boiled monkey.

"When we're down to eating our ancestors," she asked, "what is left?"

So far from the forests and savannas of our origins, few of us still sense a link to our animal forebears. That the Amazonian Zápara actually do is remarkable, since the divergence of humans from other primates occurred on another continent. Nevertheless, lately we have had a creeping sense of what Ana María means. Even if we're not driven to cannibalism, might we, too, face terrible choices as we skulk toward the future?

A generation ago, humans eluded nuclear annihilation; with luck, we'll continue to dodge that and other mass terrors. But now we often find ourselves asking whether inadvertently we've poisoned or parboiled the planet, ourselves included. We've also used and abused water and soil so that there's a lot less of each, and trampled thousands of species that probably aren't coming back. Our world, some respected voices warn, could one day degenerate into something resembling a vacant lot, where crows and rats scuttle among weeds, preying on each other. If it comes to that, at what point would things have gone so far that, for all our vaunted superior intelligence, we're not among the hardy survivors?

The truth is, we don't know. Any conjecture gets muddled by our obstinate reluctance to accept that the worst might actually occur. We may be undermined by our survival instincts, honed over eons to help us deny, defy, or ignore catastrophic portents lest they paralyze us with fright.

If those instincts dupe us into waiting until it's too late, that's bad. If they fortify our resistance in the face of mounting omens, that's good. More than once, crazy, stubborn hope has inspired creative strokes that snatched people from ruin. So, let us try a creative experiment: Suppose that the worst has happened. Human extinction is a fait accompli. Not by nuclear calamity, asteroid collision, or anything ruinous enough to also wipe out most everything else, leaving whatever remained in some radically altered, reduced state. Nor by some grim eco-scenario in which we agonizingly fade, dragging many more species with us in the process.

Instead, picture a world from which we all suddenly vanished. Tomorrow.

Unlikely perhaps, but for the sake of argument, not impossible. Say a *Homo sapiens*–specific virus—natural or diabolically nano-engineered—picks us off but leaves everything else intact. Or some misanthropic evil wizard somehow targets that unique 3.9 percent of DNA that makes us human beings and not chimpanzees, or perfects a way to sterilize our sperm. Or say that Jesus—more on Him later—or space aliens rapture us away, either to our heavenly glory or to a zoo somewhere across the galaxy.

Look around you, at today's world. Your house, your city. The surrounding land, the pavement underneath, and the soil hidden below that. Leave it all in place, but extract the human beings. Wipe us out, and see what's left. How would the rest of nature respond if it were suddenly relieved of the relentless pressures we heap on it and our fellow organisms? How soon would, or could, the climate return to where it was before we fired up all our engines?

How long would it take to recover lost ground and restore Eden to the way it must have gleamed and smelled the day before Adam, or *Homo habilis,* appeared? Could nature ever obliterate all our traces? How would it undo our monumental cities and public works, and reduce our myriad plastics and toxic synthetics back to benign, basic elements? Or are some so unnatural that they're indestructible?

And what of our finest creations—our architecture, our art, our many manifestations of spirit? Are any truly timeless, at least enough so to last until the sun expands and roasts our Earth to a cinder?

And even after *that,* might we have left some faint, enduring mark on the universe; some lasting glow, or echo, of Earthly humanity; some interplanetary sign that once we were here?

For a sense of how the world would go on without us, among other places we must look to the world before us. We're not time travelers, and the fossil record is only a fragmentary sampling. But even if that record were complete, the future won't perfectly mirror the past. We've ground some species so thoroughly into extinction that they, or their DNA, will likely never spring back. Since some things we've done are likely irrevocable, what would remain in our absence would not be the same planet had we never evolved in the first place.

Yet it might not be so different, either. Nature has been through worse losses before, and refilled empty niches. And even today, there are still a few Earthly spots where all our senses can inhale a living memory of this Eden before we were here. Inevitably they invite us to wonder how nature might flourish if granted the chance.

Since we're imagining, why not also dream of a way for nature to prosper that doesn't depend on our demise? We are, after all, mammals ourselves. Every life-form adds to this vast pageant. With our passing, might some lost contribution of ours leave the planet a bit more impoverished?

Is it possible that, instead of heaving a huge biological sigh of relief, the world without us would miss us?

PART I

CHAPTER 1

❧

A Lingering Scent of Eden

YOU MAY NEVER have heard of the Białowieża Puszcza. But if you were raised somewhere in the temperate swathe that crosses much of North America, Japan, Korea, Russia, several former Soviet republics, parts of China, Turkey, and Eastern and Western Europe—including the British Isles—something within you remembers it. If instead you were born to tundra or desert, subtropics or tropics, pampas or savannas, there are still places on Earth kindred to this *puszcza* to stir your memory, too.

Puszcza, an old Polish word, means "forest primeval." Straddling the border between Poland and Belarus, the half-million acres of the Białowieża Puszcza contain Europe's last remaining fragment of old-growth, lowland wilderness. Think of the misty, brooding forest that loomed behind your eyelids when, as a child, someone read you the Grimm Brothers' fairy tales. Here, ash and linden trees tower nearly 150 feet, their huge canopies shading a moist, tangled understory of hornbeams, ferns, swamp alders and crockery-sized fungi. Oaks, shrouded with half a millennium of moss, grow so immense here that great spotted woodpeckers store spruce cones in their three-inch-deep bark furrows. The air, thick and cool, is draped with silence that parts briefly for a nutcracker's croak, a pygmy owl's low whistle, or a wolf's wail, then returns to stillness.

The fragrance that wafts from eons of accumulated mulch in the forest's core hearkens to fertility's very origins. In the Białowieża, the profusion of life owes much to all that is dead. Almost a quarter of the organic mass aboveground is in assorted stages of decay—more than 50 cubic yards of decomposing trunks and fallen branches on every acre, nourishing thousands of species of mushrooms, lichens, bark beetles, grubs, and microbes that are missing from the orderly, managed woodlands that pass as forests elsewhere.

Together those species stock a sylvan larder that provides for weasels, pine martens, raccoons, badgers, otters, fox, lynx, wolves, roe deer, elk, and eagles. More kinds of life are found here than anywhere else on the continent—yet there are no surrounding mountains or sheltering valleys to form unique niches for endemic species. The Białowieża Puszcza is simply a relic of what once stretched east to Siberia and west to Ireland.

The existence in Europe of such a legacy of unbroken biological antiquity owes, unsurprisingly, to high privilege. During the 14th century, a Lithuanian duke named Władysław Jagiełło, having successfully allied his grand duchy with the Kingdom of Poland, declared the forest a royal hunting preserve. For centuries, it stayed that way. When the Polish-Lithuanian union was finally subsumed by Russia, the Białowieża became the private domain of the tsars. Although occupying Germans took lumber and slaughtered game during World War I, a pristine core was left intact, which in 1921 became a Polish national park. The timber pillaging resumed briefly under the Soviets, but when the Nazis invaded, a nature fanatic named Hermann Göring decreed the entire preserve off-limits, except by his pleasure.

Following World War II, a reportedly drunken Josef Stalin agreed one evening in Warsaw to let Poland retain two-fifths of the forest. Little else changed under communist rule, except for construction of some elite hunting dachas—in one of which, Viskuli, an agreement was signed in 1991 dissolving the Soviet Union into free states. Yet, as it turns out, this ancient sanctuary is more threatened under Polish democracy and Belarusian independence than it was during seven centuries of monarchs and dictators. Forestry ministries in both countries tout increased management to preserve the Puszcza's health. Management, however, often turns out to be a euphemism for culling—and selling—mature hardwoods that otherwise would one day return a windfall of nutrients to the forest.

≈

IT IS STARTLING to think that all Europe once looked like this Puszcza. To enter it is to realize that most of us were bred to a pale copy of what

Five-hundred-year-old oaks. Białowieża Puszcza, Poland.

PHOTO BY JANUSZ KORBEL.

nature intended. Seeing elders with trunks seven feet wide, or walking through stands of the tallest trees here—gigantic Norway spruce, shaggy as Methuselah—should seem as exotic as the Amazon or Antarctica to someone raised among the comparatively puny, second-growth woodlands found throughout the Northern Hemisphere. Instead, what's astonishing is how primally familiar it feels. And, on some cellular level, how complete.

Andrzej Bobiec recognized it instantly. As a forestry student in Krakow, he'd been trained to manage forests for maximum productivity, which included removing "excess" organic litter lest it harbor pests like bark beetles. Then, on a visit here he was stunned to discover 10 times more biodiversity than in any forest he'd ever seen.

It was the only place left with all nine European woodpecker species, because, he realized, some of them only nest in hollow, dying trees. "They can't survive in managed forests," he argued to his forestry professors. "The Białowieża Puszcza has managed itself perfectly well for millennia."

The husky, bearded young Polish forester became instead a forest ecologist. He was hired by the Polish national park service. Eventually, he was fired for protesting management plans that chipped ever closer to the pristine core of the Puszcza. In various international journals, he blistered official policies that asserted that "forests will die without our thoughtful help," or that justified cutting timber in the Białowieża's surrounding buffer to "reestablish the primeval character of stands." Such convoluted thinking, he accused, was rampant among Europeans who have hardly any memory of forested wilderness.

To keep his own memory connected, for years he daily laced his leather boots and hiked through his beloved Puszcza. Yet although he ferociously defends those parts of this forest still undisturbed by man, Andrzej Bobiec can't help being seduced by his own human nature.

Alone in the woods, Bobiec enters into communion with fellow *Homo sapiens* through the ages. A wilderness this pure is a blank slate to record human passage: a record he has learned to read. Charcoal layers in the soil show him where gamesmen once used fire to clear parts of the forest for browse. Stands of birch and trembling aspen attest to a time when Jagiełło's descendants were distracted from hunting, perhaps by war, long enough for these sun-seeking species to recolonize game clearings. In their shade grow telltale seedlings of the hardwoods that were here before them.

Gradually, these will crowd out the birch and aspen, until it will be as if they were never gone.

Whenever Bobiec happens on an anomalous shrub like hawthorn or on an old apple tree, he knows he's in the presence of the ghost of a log house long ago devoured by the same microbes that can turn the giant trees here back into soil. Any lone, massive oak he finds growing from a low, clover-covered mound marks a crematorium. Its roots have drawn nourishment from the ashes of Slavic ancestors of today's Belorusians, who came from the east 900 years ago. On the northwest edge of the forest, Jews from five surrounding shtetls buried their dead. Their sandstone and granite headstones from the 1850s, mossy and tumbled by roots, have already worn so smooth that they've begun to resemble the pebbles left by their mourning relatives, who themselves long ago departed.

Andrzej Bobiec passes through a blue-green glade of Scots pine, barely a mile from the Belarusian border. The waning October afternoon is so hushed, he can hear snowflakes alight. Suddenly, there's a crashing in the underbrush, and a dozen wisent—*Bison bonasus,* European bison—burst from where they've been browsing on young shoots. Steaming and pawing, their huge black eyes glance just long enough for them to do what their own ancestors discovered they must upon encountering one of these deceptively frail bipeds: they flee.

Just 600 wisent remain in the wild, nearly all of them here—or just half, depending on what's meant by *here.* An iron curtain bisects this paradise, erected by the Soviets in 1980 along the border to thwart escapees to Poland's renegade Solidarity movement. Although wolves dig under it, and roe deer and elk are believed to leap it, the herd of these largest of Europe's mammals remains divided, and with it, its gene pool—divided and mortally diminished, some zoologists fear. Once, following World War I, bison from zoos were brought here to replenish a species nearly extirpated by hungry soldiers. Now, a remnant of a Cold War threatens them again.

Belarus, which well after communism's collapse has yet to remove statues of Lenin, also shows no inclination to dismantle the fence, especially as Poland's border is now the European Union's. Although just 14 kilometers separate the two countries' park headquarters, to see the Belovezhskaya Pushcha, as it is called in Belorusian, a foreign visitor must drive 100 miles south, take a train across the border to the city of Brest, submit to pointless interrogation, and hire a car to drive back north. Andrzej Bobiec's Belorusian counterpart and fellow activist, Heorhi Kazulka, is a pale, sallow invertebrate biologist and former deputy director of Belarus's side of the primeval forest. He was also fired by his own country's park service, for challenging one of the latest park additions—a sawmill. He cannot risk being seen with Westerners. Inside the Brezhnev-era tenement where he lives at the forest's edge, he apologetically offers visitors tea and discusses his dream of an international peace park where bison and moose would roam and breed freely.

The Pushcha's colossal trees are the same as those in Poland; the same buttercups, lichens, and enormous red oak leaves; the same circling white-tailed eagles, heedless of the razor-wire barrier below. In fact, on both sides, the forest is actually growing, as peasant populations leave shrinking villages for cities. In this moist climate, birch and aspen quickly invade their fallow potato fields; within just two decades, farmland gives way to woodland. Under the canopy of the pioneering trees, oak, maple, linden, elm, and spruce regenerate. Given 500 years without people, a true forest could return.

The thought of rural Europe reverting one day to original forest is heartening. But unless the last humans remember to first remove Belarus's iron curtain, its bison may wither away with them.

❧

CHAPTER 2

Unbuilding Our Home

" 'If you want to destroy a barn,' a farmer once told me,
'cut an eighteen-inch-square hole in the roof.
Then stand back.' "

—architect Chris Riddle
Amherst, Massachusetts

O N T H E D A Y after humans disappear, nature takes over and immediately begins cleaning house—or houses, that is. Cleans them right off the face of the Earth. They all go.

If you're a homeowner, you already knew it was only a matter of time for yours, but you've resisted admitting it, even as erosion callously attacked, starting with your savings. Back when they told you what your house would cost, nobody mentioned what you'd also be paying so that nature wouldn't repossess it long before the bank.

Even if you live in a denatured, postmodern subdivision where heavy machines mashed the landscape into submission, replacing unruly native flora with obedient sod and uniform saplings, and paving wetlands in the righteous name of mosquito control—even then, you know that nature wasn't fazed. No matter how hermetically you've sealed your temperature-tuned interior from the weather, invisible spores penetrate anyway, exploding in sudden outbursts of mold—awful when you see it, worse when you don't, because it's hidden behind a painted wall, munching paper sandwiches of gypsum board, rotting studs and floor joists. Or you've been colonized by termites, carpenter ants, roaches, hornets, even small mammals.

Most of all, though, you are beset by what in other contexts is the veritable stuff of life: water. It always wants in.

After we're gone, nature's revenge for our smug, mechanized superiority arrives waterborne. It starts with wood-frame construction, the most widely used residential building technique in the developed world. It begins on the roof, probably asphalt or slate shingle, warranted to last two or three decades—but that warranty doesn't count around the chimney, where the first leak occurs. As the flashing separates under rain's relentless insistence, water sneaks beneath the shingles. It flows across four-by-eight-foot sheets of sheathing made either of plywood or, if newer, of woodchip board composed of three-to four-inch flakes of timber, bonded together by a resin.

Newer isn't necessarily better. Wernher Von Braun, the German scientist who developed the U.S. space program, used to tell a story about Colonel John Glenn, the first American to orbit the Earth. "Seconds before lift-off, with Glenn strapped into that rocket we built for him and man's best efforts all focused on that moment, you know what he said to himself? 'Oh, my God! I'm sitting on a pile of low bids!'"

In your new house, you've been sitting under one. On the one hand, that's all right: by building things so cheaply and lightly, we use fewer of the world's resources. On the other hand, the massive trees that yielded the great wooden posts and beams that still support medieval European, Japanese, and early American walls are now too precious and rare, and we're left to make do with gluing together smaller boards and scraps.

The resin in your cost-conscious choice of a woodchip roof, a waterproof goo of formaldehyde and phenol polymer, was also applied along the board's exposed edges, but it fails anyway because moisture enters around the nails. Soon they're rusting, and their grip begins to loosen. That presently leads not only to interior leaks, but to structural mayhem. Besides underlying the roofing, the wooden sheathing secures trusses to each other. The trusses—premanufactured braces held together with metal connection plates—are there to keep the roof from splaying. But when the sheathing goes, structural integrity goes with it.

As gravity increases tension on the trusses, the ¼-inch pins securing their now-rusting connector plates pull free from the wet wood, which now sports a fuzzy coating of greenish mold. Beneath the mold, threadlike filaments called hyphae are secreting enzymes that break cellulose and lignin down into fungi food. The same thing is happening to the floors inside. When the heat went off, pipes burst if you lived where it freezes, and rain is blowing in where windows have cracked from bird collisions and the stress of sagging walls. Even where the glass is still intact, rain and snow mysteriously, inexorably work their way under sills. As the wood continues to rot, trusses start to collapse against each other. Eventually the walls lean to one side, and finally the roof falls in. That barn roof with the 18-by-18-inch hole was likely gone inside of 10 years. Your house's lasts maybe 50 years; 100, tops.

While all that disaster was unfolding, squirrels, raccoons, and lizards have been inside, chewing nest holes in the drywall, even as woodpeckers rammed their way through from the other direction. If they were initially thwarted by allegedly indestructible siding made of aluminum, vinyl, or the maintenance-free, portland-cement-cellulose-fiber clapboards known as Hardie planks, they merely have to wait a century before most of it is lying on the ground. Its factory-impregnated color is nearly gone, and as water works its inevitable way into saw cuts and holes where the planks took nails, bacteria are picking over its vegetable matter and leaving its minerals behind. Fallen vinyl siding, whose color began to fade early, is now brittle and cracking as its plasticizers degenerate. The aluminum is in better shape, but salts in water pooling on its surface slowly eat little pits that leave a grainy white coating.

For many decades, even after being exposed to the elements, zinc galvanizing has protected your steel heating and cooling ducts. But water and air have been conspiring to convert it to zinc oxide. Once the coating is consumed, the unprotected thin sheet steel disintegrates in a few years. Long before that, the water-soluble gypsum in the sheetrock has washed back into the earth. That leaves the chimney, where all the trouble began. After a century, it's still standing, but its bricks have begun to drop and break as, little by little, its lime mortar, exposed to temperature swings, crumbles and powders.

If you owned a swimming pool, it's now a planter box, filled with either the offspring of ornamental saplings that the developer imported, or with banished natural foliage that was still hovering on the subdivision's fringes, awaiting the chance to retake its territory. If the house's foundation involved a basement, it too is filling with soil and plant life. Brambles and wild grapevines are snaking around steel gas pipes, which will rust away before another century goes by. White plastic PVC plumbing has yellowed and thinned on the side exposed to the light, where its chloride is weathering to hydrochloric acid, dissolving itself and its polyvinyl part-ners. Only the bathroom tile, the chemical properties of its fired ceramic not unlike those of fossils, is relatively unchanged, although it now lies in a pile mixed with leaf litter.

After 500 years, what is left depends on where in the world you lived. If the climate was temperate, a forest stands in place of a suburb; minus a few hills, it's begun to resemble what it was before developers, or the farm-ers they expropriated, first saw it. Amid the trees, half-concealed by a spreading understory, lie aluminum dishwasher parts and stainless steel cookware, their plastic handles splitting but still solid. Over the coming centuries, although there will be no metallurgists around to measure it, the pace at which aluminum pits and corrodes will finally be revealed: a rela-tively new material, aluminum was unknown to early humans because its ore must be electrochemically refined to form metal.

The chromium alloys that give stainless steel its resilience, however, will probably continue to do so for millennia, especially if the pots, pans, and carbon-tempered cutlery are buried out of the reach of atmospheric oxygen. One hundred thousand years hence, the intellectual development of what-ever creature digs them up might be kicked abruptly to a higher evolutionary plane by the discovery of ready-made tools. Then again, lack of knowledge of how to duplicate them could be a demoralizing frustration—or an awe-arousing mystery that ignites religious consciousness.

If you were a desert dweller, the plastic components of modern life flake and peel away faster, as polymer chains crack under an ultraviolet barrage of daily sunshine. With less moisture, wood lasts longer there, though any metal in contact with salty desert soils will corrode more quickly. Still, from Roman ruins we can guess that thick cast iron will be around well into the future's archaeological record, so the odd prospect of fire hydrants sprouting amidst cacti may someday be among the few clues that humanity was here. Although adobe and plaster walls will have eroded away, the wrought iron balconies and window grates that once adorned them may still be recognizable, albeit airy as tulle, as corrosion eating through the iron encounters its matrix of indigestible glass slag.

Once, we built structures entirely from the most durable substances we knew: granite block, for instance. The results are still around today to admire, but we don't often emulate them, because quarrying, cutting, transporting, and fitting stone require a patience we no longer possess. No one since the likes of Antoni Gaudí, who began Barcelona's yet-unfinished Sagrada Familia basilica in 1880, contemplates investing in construction that our great-great-grandchildren's grandchildren will complete 250 years hence. Nor, absent the availability of a few thousand slaves, is it cheap, especially compared to another Roman innovation: concrete.

Today, that brew of clay, sand, and a paste made of the calcium of ancient seashells hardens into a man-made rock that is increasingly the most affordable option for *Homo sapiens urbanus*. What happens, then, to the cement cities now home to more than half the humans alive?

Before we consider that, there's a matter to address regarding climate. If we were to vanish tomorrow, the momentum of certain forces we've already set in motion will continue until centuries of gravity, chemistry, and entropy slow them to an equilibrium that may only partly resemble the one that existed before us. That former equilibrium depended on a sizeable amount of carbon locked away beneath Earth's crust, much of which we've now relocated into the atmosphere. Instead of rotting, the wood frames of houses may be preserved like the timbers of Spanish galleons wherever rising seas pickle them in salt water.

In a warmer world, the deserts may grow drier, but the parts where humans dwelled will likely again be visited by what attracted those humans in the first place: flowing water. From Cairo to Phoenix, desert cities rose where rivers made arid soils livable. Then, as population grew, humans seized control of those aquatic arteries, diverting them in ways that allowed for even more growth. But after people are gone, the diversions will soon follow them. Drier, hotter desert climates will be complemented by wetter, stormier mountain weather systems that will send floods roaring downstream, overwhelming dams, spreading over their former alluvial plains, and entombing whatever was built there in annual layers of silt. Within them, fire hydrants, truck tires, shattered plate glass, condominia, and office buildings may remain indefinitely, but as far from sight as the Carboniferous Formation once was.

No memorial will mark their burial, though the roots of cottonwoods, willows, and palms may occasionally make note of their presence. Only eons later, when old mountains have worn away and new ones risen, will young streams cutting fresh canyons through sediments reveal what once, briefly, went on here.

The City Without Us

THE NOTION THAT someday nature could swallow whole something
so colossal and concrete as a modern city doesn't slide easily into our
imaginations. The sheer titanic presence of a New York City resists efforts
to picture it wasting away. The events of September 2001 showed only
what human beings with explosive hardware can do, not crude processes
like erosion or rot. The breathtaking, swift collapse of the World Trade
Center towers suggested more to us about their attackers than about mortal vulnerabilities that could doom our entire infrastructure. And even that
once-inconceivable calamity was confined to just a few buildings. Nevertheless, the time it would take nature to rid itself of what urbanity has
wrought may be less than we might suspect.

ودو

IN 1939, A World's Fair was held in New York. For its exhibit, the government of Poland sent a statue of Władysław Jagiełło. The founder of
the Białowieża Puszcza had not been immortalized in bronze for preserving a chunk of primeval forest six centuries earlier. By marrying its queen,
Jagiełło had united Poland and his duchy of Lithuania into a European
power. The sculpture portrays him on horseback following his victory at
the Battle of Grünwald in 1410. Triumphant, he hoists two swords captured from Poland's latest vanquished enemy, the Teutonic Knights of the ·
Cross.

In 1939, however, the Poles weren't faring so well against some descendants of those Teutonic Knights. Before the New York World's Fair ended, Hitler's Nazis had taken Poland, and the sculpture couldn't be returned to its homeland. Six sad years later, the Polish government gave it to New York as a symbol of its courageous, battered survivors. The statue of Jagiełło was placed in Central Park, overlooking what today is called Turtle Pond.

When Dr. Eric Sanderson leads a tour through the park, he and his flock usually pass Jagiełło without pausing, because they are lost in another century altogether—the 17th. Bespectacled under his wide-brimmed felt hat, a trim beard graying around his chin and a laptop jammed in his backpack, Sanderson is a landscape ecologist with the Wildlife Conservation Society, a global squadron of researchers trying to save an imperiled world from itself. At its Bronx Zoo headquarters, Sanderson directs the Mannahatta Project, an attempt to re-create, virtually, Manhattan Island as it was when Henry Hudson's crew first saw it in 1609: a pre-urban vision that tempts speculation about how a posthuman future might look.

His team has scoured original Dutch documents, colonial British military maps, topographic surveys, and centuries of assorted archives throughout town. They've probed sediments, analyzed fossil pollens, and plugged thousands of bits of biological data into imaging software that generates three-dimensional panoramas of the heavily wooded wilderness on which a metropolis was juxtaposed. With each new entry of a species of grass or tree that is historically confirmed in some part of the city, the images grow more detailed, more startling, more convincing. Their goal is a block-by-city-block guide to this ghost forest, the one Eric Sanderson uncannily seems to see even while dodging Fifth Avenue buses.

When Sanderson wanders through Central Park, he's able to look beyond the half-million cubic yards of soil hauled in by its designers, Frederick Law Olmstead and Calvert Vaux, to fill in what was mostly a swampy bog surrounded by poison oak and sumac. He can trace the shoreline of the long, narrow lake that lay along what is now 59th Street, north of the Plaza Hotel, with its tidal outlet that meandered through salt marsh to the East River. From the west, he can see a pair of streams entering the lake that drained the slope of Manhattan's major ridgeline, a deer and mountain lion trail known today as Broadway.

Eric Sanderson sees water flowing everywhere in town, much of it bubbling from underground ("which is how Spring Street got its name"). He's

Manhattan, circa 1609, juxtaposed with Manhattan, circa 2006, showing
infilling that has extended the island's southern tip.

© YANN ARTHUS-BERTRAND/CORBIS; 3D VISUALIZATION BY MARKLEY BOYER FOR
THE MANNAHATTA PROJECT/WILDLIFE CONSERVATION SOCIETY.

identified more than 40 brooks and streams that traversed what was once a
hilly, rocky island: in the Algonquin tongue of its first human occupants, the
Lenni Lenape, *Mannahatta* referred to those now-vanished hills. When New
York's 19th-century planners imposed a grid on everything north of Greenwich Village—the jumble of original streets to the south being impossible
to unsnarl—they behaved as if topography were irrelevant. Except for
some massive, unmoveable schist outcrops in Central Park and at the island's
northern tip, Manhattan's textured terrain was squashed and dumped into
streambeds, then planed and leveled to receive the advancing city.

Later, new contours arose, this time routed through rectilinear forms
and hard angles, much as the water that once sculpted the island's land was
now forced underground through a lattice of pipes. Eric Sanderson's
Mannahatta Project has plotted how closely the modern sewer system follows the old watercourses, although man-made sewer lines can't wick away
runoff as efficiently as nature. In a city that buried its rivers, he observes,
"rain still falls. It has to go somewhere."

As it happens, that will be the key to breaching Manhattan's hard shell if nature sets about dismantling it. It would begin very quickly, with the first strike at the city's most vulnerable spot: its underbelly.

New York City Transit's Paul Schuber and Peter Briffa, superintendent of Hydraulics and level one maintenance supervisor of Hydraulics Emergency Response, respectively, understand perfectly how this would work. Every day, they must keep 13 million gallons of water from overpowering New York's subway tunnels.

"That's just the water that's already underground," notes Schuber.

"When it rains, the amount is . . ." Briffa shows his palms, surrendering. "It's incalculable."

Maybe not actually incalculable, but it doesn't rain any less now than before the city was built. Once, Manhattan was 27 square miles of porous ground interlaced with living roots that siphoned the 47.2 inches of average annual rainfall up trees and into meadow grasses, which drank their fill and exhaled the rest back into the atmosphere. Whatever the roots didn't take settled into the island's water table. In places, it surfaced in lakes and marshes, with the excess draining off to the ocean via those 40 streams—which now lie trapped beneath concrete and asphalt.

Today, because there's little soil to absorb rainfall or vegetation to transpire it, and because buildings block sunlight from evaporating it, rain collects in puddles or follows gravity down sewers—or it flows into subway vents, adding to the water already down there. Below 131st Street and Lenox Avenue, for example, a rising underground river is corroding the bottom of the A, B, C, and D subway lines. Constantly, men in reflective vests and denim rough-outs like Schuber's and Briffa's are clambering around beneath the city to deal with the fact that under New York, groundwater is always rising.

Whenever it rains hard, sewers clog with storm debris—the number of plastic garbage bags adrift in the world's cities may truly exceed calculation—and the water, needing to go somewhere, plops down the nearest subway stairs. Add a nor'easter, and the surging Atlantic Ocean bangs against New York's water table until, in places like Water Street in lower Manhattan or Yankee Stadium in the Bronx, it backs up right into the tunnels, shutting everything down until it subsides. Should the ocean continue to warm and rise even faster than the current inch per decade, at some point it simply won't subside. Schuber and Briffa have no idea what will happen then.

Add to all that the 1930s-vintage water mains that frequently burst, and the only thing that has kept New York from flooding already is the incessant vigilance of its subway crews and 753 pumps. Think about those pumps: New York's subway system, an engineering marvel in 1903, was laid underneath an already-existing, burgeoning city. As that city already had sewer lines, the only place for subways to go was below them. "So," explains Schuber, "we have to pump uphill." In this, New York is not alone: cities like London, Moscow, and Washington built their subways far deeper, often to double as bomb shelters. Therein lies much potential disaster.

Shading his eyes with his white hard hat, Schuber peers down into a square pit beneath the Van Siclen Avenue station in Brooklyn, where each minute 650 gallons of natural groundwater gush from the bedrock. Gesturing over the roaring cascade, he indicates four submersible cast-iron pumps that take turns laboring against gravity to stay ahead. Such pumps run on electricity. When the power fails, things can get difficult very fast. Following the World Trade Center attack, an emergency pump train bearing a jumbo portable diesel generator pumped out 27 times the volume of Shea Stadium. Had the Hudson River actually burst through the PATH train tunnels that connect New York's subways to New Jersey, as was greatly feared, the pump train—and possibly much of the city—would simply have been overwhelmed.

In an abandoned city, there would be no one like Paul Schuber and Peter Briffa to race from station to flooded station whenever more than two inches of rain falls—as happens lately with disturbing frequency—sometimes snaking hoses up stairways to pump to a sewer down the street, sometimes navigating these tunnels in inflatable boats. With no people, there would also be no power. The pumps will go off, and stay off. "When this pump facility shuts down," says Schuber, "in half an hour water reaches a level where trains can't pass anymore."

Briffa removes his safety goggles and rubs his eyes. "A flood in one zone would push water into the others. Within 36 hours, the whole thing could fill."

Even if it weren't raining, with subway pumps stilled, that would take no more than a couple of days, they estimate. At that point, water would start sluicing away soil under the pavement. Before long, streets start to crater. With no one unclogging sewers, some new watercourses form on the surface. Others appear suddenly as waterlogged subway ceilings collapse. Within 20 years, the water-soaked steel columns that support the street above the East Side's 4, 5, and 6 trains corrode and buckle. As Lexington Avenue caves in, it becomes a river.

Well before then, however, pavement all over town would have already been in trouble. According to Dr. Jameel Ahmad, chairman of the civil engineering department at New York's Cooper Union, things will begin to fall apart during the first month of March after humans vacate Manhattan. Each March, temperatures normally flutter back and forth around 32°F as many as 40 times (presumably, climate change could push this back to February). Whenever it is, the repeated freezing and thawing make asphalt and cement split. When snow thaws, water seeps into these fresh cracks. When it freezes, the water expands, and cracks widen.

Call it water's retaliation for being squished under all that cityscape. Almost every other compound in nature contracts when frozen, but H_2O molecules do the opposite, organizing themselves into elegant hexagonal crystals that take up about 9 percent more space than they did when sloshing around in a liquid state. Pretty six-sided crystals suggest snowflakes so gossamer it's hard to conceive of them pushing apart slabs of sidewalk. It's even more difficult to imagine carbon steel water pipes built to withstand 7,500 pounds of pressure per square inch exploding when they freeze. Yet that's exactly what happens.

As pavement separates, weeds like mustard, shamrock, and goosegrass blow in from Central Park and work their way down the new cracks, which widen further. In the current world, before they get too far, city maintenance usually shows up, kills the weeds, and fills the fissures. But in the post-people world, there's no one left to continually patch New York. The weeds are followed by the city's most prolific exotic species, the Chinese ailanthus tree. Even with 8 million people around, ailanthus—otherwise innocently known as the tree-of-heaven—are implacable invaders capable of rooting in tiny chinks in subway tunnels, unnoticed until their spreading leaf canopies start poking from sidewalk grates. With no one to yank their seedlings, within five years powerful ailanthus roots are heaving up sidewalks and wreaking havoc in sewers—which are already stressed by all the plastic bags and old newspaper mush that no one is clearing away. As soil long trapped beneath pavement gets exposed to sun and rain, other species jump in, and soon leaf litter adds to the rising piles of debris clogging the sewer grates.

The early pioneer plants won't even have to wait for the pavement to fall apart. Starting from the mulch collecting in gutters, a layer of soil will start forming atop New York's sterile hard shell, and seedlings will sprout. With far less organic material available to it—just windblown dust and urban soot—precisely that has happened in an abandoned elevated iron bed of the New York Central Railroad on Manhattan's West Side. Since trains stopped running there in 1980, the inevitable ailanthus trees have been joined by a thickening ground cover of onion grass and fuzzy lamb's ear, accented by stands of goldenrod. In some places, the track emerges from the second stories of warehouses it once serviced into elevated lanes of wild crocuses, irises, evening primrose, asters, and Queen Anne's lace. So many New Yorkers, glancing down from windows in Chelsea's art district, were moved by the sight of this untended, flowering green ribbon, prophetically and swiftly laying claim to a dead slice of their city, that it was dubbed the High Line and officially designated a park.

In the first few years with no heat, pipes burst all over town, the freeze-thaw cycle moves indoors, and things start to seriously deteriorate. Buildings groan as their innards expand and contract; joints between walls and rooflines separate. Where they do, rain leaks in, bolts rust, and facing pops off, exposing insulation. If the city hasn't burned yet, it will now. Collectively, New York architecture isn't as combustible as, say, San Francisco's incendiary rows of clapboard Victorians. But with no firemen to answer the call, a dry lightning strike that ignites a decade of dead branches and leaves piling up in Central Park will spread flames through the streets. Within two decades, lightning rods have begun to rust and snap, and roof fires leap among buildings, entering paneled offices filled with paper fuel. Gas lines ignite with a rush of flames that blows out windows. Rain and snow blow in, and soon even poured concrete floors are freezing, thawing, and starting to buckle. Burnt insulation and charred wood add nutrients to Manhattan's growing soil cap. Native Virginia creeper and poison ivy claw at walls covered with lichens, which thrive in the absence of air pollution.

Red-tailed hawks and peregrine falcons nest in increasingly skeletal high-rise structures.

Within two centuries, estimates Brooklyn Botanical Garden vice president Steven Clemants, colonizing trees will have substantially replaced pioneer weeds. Gutters buried under tons of leaf litter provide new, fertile ground for native oaks and maples from city parks. Arriving black locust and autumn olive shrubs fix nitrogen, allowing sunflowers, bluestem, and white snakeroot to move in along with apple trees, their seeds expelled by proliferating birds.

Biodiversity will increase even more, predicts Cooper Union civil engineering chair Jameel Ahmad, as buildings tumble and smash into each other, and lime from crushed concrete raises soil pH, inviting in trees, such as buckthorn and birch, that need less-acidic environments. Ahmad, a hearty silver-haired man whose hands talk in descriptive circles, believes that process will begin faster than people might think. A native of Lahore, Pakistan, a city of ancient mosaic-encrusted mosques, he now teaches how to design and retrofit buildings to withstand terrorist attacks, and has accrued a keen understanding of structural weakness.

"Even buildings anchored into hard Manhattan schist, like most New York skyscrapers," he notes, "weren't intended to have their steel foundations waterlogged." Plugged sewers, deluged tunnels, and streets reverting to rivers, he says, will conspire to undermine subbasements and destabilize their huge loads. In a future that portends stronger and more-frequent hurricanes striking North America's Atlantic coast, ferocious winds will pummel tall, unsteady structures. Some will topple, knocking down others. Like a gap in the forest when a giant tree falls, new growth will rush in. Gradually, the asphalt jungle will give way to a real one.

⚶

The New York Botanical Garden, located on 250 acres across from the Bronx Zoo, possesses the largest herbarium anywhere outside of Europe. Among its treasures are wildflower specimens gathered on Captain Cook's 1769 Pacific wanderings, and a shred of moss from Tierra del Fuego, with accompanying notes written in watery black ink and signed by its collector, C. Darwin. Most remarkable, though, is the NYBG's 40-acre tract of original, old-growth, virgin New York forest, never logged.

Never cut, but mightily changed. Until only recently, it was known as the Hemlock Forest for its shady stands of that graceful conifer, but almost every hemlock here is now dead, slain by a Japanese insect smaller than the period at the end of this sentence, which arrived in New York in the mid-1980s. The oldest and biggest oaks, dating back to when this forest was British, are also crashing down, their vigor sapped by acid rain and heavy metals such as lead from automobile and factory fumes, which have soaked into the soil. It's unlikely that they'll come back, because most canopy trees here long ago stopped regenerating. Every resident native species now harbors its own pathogen: some fungus, insect, or disease that seizes the opportunity to ravish trees weakened by chemical onslaught. As if that weren't enough, as the NYBG forest became an island of greenery surrounded by hundreds of square miles of gray urbanity, it became the primary refuge for Bronx squirrels. With natural predators gone and no hunting permitted, there's nothing to stop them from devouring every acorn or hickory nut before it can germinate. Which they do.

There is now an eight-decade gap in this old forest's understory. Instead of new generations of native oaks, maple, ash, birch, sycamore and tulip trees, what's mainly growing are imported ornamentals that have blown in from the rest of the Bronx. Soil samplings indicate some 20 million ailanthus seeds sprouting here. According to Chuck Peters, curator of the NYBG's Institute of Economic Botany, exotics such as ailanthus and cork trees, both from China, now account for more than a quarter of this forest.

"Some people want to put the forest back the way it was 200 years ago," he says. "To do that, I tell them, you've got to put the Bronx back the way it was 200 years ago."

As human beings learned to transport themselves all over the world, they took living things with them and brought back others. Plants from the Americas changed not only ecosystems in European countries but also their very identities: think of Ireland before potatoes, or Italy before tomatoes. In the opposite direction, Old World invaders not only forced themselves on hapless women of vanquished new lands, but broadcast other kinds of seed, beginning with wheat, barley, and rye. In a phrase coined by the American geographer Alfred Crosby, this ecological imperialism helped European conquerors to permanently stamp their image on their colonies.

Some results were ludicrous, like English gardens with hyacinths and daffodils that never quite took hold in colonial India. In New York, the European starling—now a ubiquitous avian pest from Alaska to Mexico—was introduced because someone thought the city would be more cultured if Central Park were home to each bird mentioned in Shakespeare. Next came a Central Park garden with every plant in the Bard's plays, sown with the lyrical likes of primrose, wormwood, lark's heel, eglantine, and cowslip—everything short of *Macbeth*'s Birnam Wood.

To what extent the Mannahatta Project's virtual past resembles the Manhattan forest to come depends on a struggle for North America's soil that will continue long after the humans that instigated it are gone. The NYBG's herbarium also holds one of the first American specimens of a deceptively lovely lavender stalk. The seeds of purple loosestrife, native to North Sea estuaries from Britain to Finland, likely arrived in wet sands that merchant ships dug from European tidal flats as ballast for the Atlantic crossing. As trade with the colonies grew, more purple loosestrife was dumped along American shores as ships jettisoned ballast before taking on cargo. Once established, it moved up streams and rivers as its seeds stuck to the muddy feathers or fur of whatever it touched. In Hudson River wetlands, communities of cattail, willow, and canary grass that fed and sheltered waterfowl and muskrats turned into solid curtains of purple, impenetrable even to wildlife. By the 21st century, purple loosestrife was at large even in Alaska, where panicked state ecologists fear it will fill entire marshes, driving out ducks, geese, terns, and swans.

Even before Shakespeare Garden, Central Park designers Olmstead and Vaux had brought in a half-million trees along with their half-million cubic yards of fill to complete their improved vision of nature, spicing up the island with exotica like Persian ironwoods, Asian katsuras, cedars of Lebanon, and Chinese royal paulownias and ginkgos. Yet once humans are gone, the native plants left to compete with a formidable contingent of alien species in order to reclaim their birthright will have some home-ground advantages.

Many foreign ornamentals—double roses, for example—will wither with the civilization that introduced them, because they are sterile hybrids that must propagate through cuttings. When the gardeners that clone them go, so do they. Other pampered colonials like English ivy, left to fend for themselves, lose to their rough American cousins, Virginia creeper and poison ivy.

Still others are really mutations, forced by highly selective breeding. If they survive at all, their form and presence will be diminished. Untended fruits such as apples—an import from Russia and Kazakhstan, belying the American Johnny Appleseed myth—select for hardiness, not appearance or taste, and turn gnarly. Except for a few survivors, unsprayed apple orchards, defenseless against their native American scourges, apple maggots and leaf miner blight, will be reclaimed by native hardwoods. Introduced garden plot vegetables will revert to their humble beginnings. Sweet carrots, originally Asian, quickly devolve to wild, unpalatable Queen Anne's lace as animals devour the last of the tasty orange ones we planted, says New York Botanical Garden vice president Dennis Stevenson. Broccoli, cabbage, Brussel sprouts, and cauliflower regress to the same unrecognizable broccoli ancestor. Descendants of seed corn planted by Dominicans in Washington Heights parkway medians may eventually retrace their DNA back to the original Mexican *teosinte,* its cob barely bigger than a sprig of wheat.

The other invasion that has accosted natives—metals such as lead, mercury, and cadmium—will not wash quickly from the soil, because these are literally heavy molecules. One thing is certain: when cars have stopped for good, and factories go dark and stay that way, no more such metals will be deposited. For the first 100 years or so, however, corrosion will periodically set off time bombs left in petroleum tanks, chemical and power plants, and hundreds of dry cleaners. Gradually, bacteria will feed on residues of fuel, laundry solvents, and lubricants, reducing them to more-benign organic hydrocarbons—although a whole spectrum of man-made novelties, ranging from certain pesticides to plasticizers to insulators, will linger for many millennia until microbes evolve to process them.

Yet with each new acid-free rainfall, trees that still endure will have fewer contaminants to resist as chemicals are gradually flushed from the system. Over centuries, vegetation will take up decreasing levels of heavy metals, and will recycle, redeposit, and dilute them further. As plants die, decay, and lay down more soil cover, the industrial toxins will be buried deeper, and each succeeding crop of native seedlings will do better.

And although many of New York's heirloom trees are endangered if not actually dying, few if any are already extinct. Even the deeply mourned American chestnut, devastated everywhere after a fungal blight entered New York around 1900 in a shipment of Asian nursery plants, still hangs on in the New York Botanical Garden's old forest—literally by its roots. It sprouts, sends up skinny shoots two feet high, gets knocked back by blight, and does it again. One day, perhaps, with no human stresses sapping its vigor, a resistant strain will finally emerge. Once the tallest hardwood in American eastern forests, the resurrected chestnut trees will have to coexist with robust non-natives that are probably here to stay—Japanese barberry, Oriental bittersweet, and surely ailanthus. The ecosystem here will be a human artifact that will persist in our absence, a cosmopolitan botanical mixture that would never have occurred without us.

Which may not be bad, suggests New York Botanical Garden's Chuck Peters. "What makes New York a great city now is its cultural diversity. Everyone has something to offer. But botanically, we're xenophobic. We love native species, and want aggressive, exotic plant species to go home."

He props his running shoe against the whitish bark of a Chinese Amur cork tree, growing among the last of the hemlocks. "This may sound blasphemous, but maintaining biodiversity is less important than maintaining a functioning ecosystem. What matters is that soil is protected, that water gets cleaned, that trees filter the air, that a canopy regenerates new seedlings to keep nutrients from draining away into the Bronx River."

He inhales a lungful of filtered Bronx air. Trim and youthful in his early fifties, Peters has spent much of his life in forests. His field research has revealed that pockets of wild palm nut trees deep in the Amazon, or of durian fruit trees in virgin Borneo, or of tea trees in Burma's jungles, aren't accidents. Once, humans were there, too. The wilderness swallowed them and their memory, but its shape still bears their echo. As will this one.

In fact, it has done so since soon after *Homo sapiens* appeared here. Eric Sanderson's Mannahatta Project is re-creating the island as the Dutch found it—not some primordial Manhattan forest no human had set foot on, because there wasn't one. "Because before the Lenni Lenape arrived," explains Sanderson, "nothing was here except for a mile-thick slab of ice."

About 11,000 years ago, as the last ice age receded northward from Manhattan, it pulled along the spruce and tamarack taiga that today grows just below the Canadian tundra. In its place came what we know as the temperate eastern forest of North America: oak, hickory, chestnut, walnut, hemlock, elm, beech, sugar maple, sweet gum, sassafras, and wild filbert. In the clearings grew shrubs of chokecherry, fragrant sumac, rhododendron, honeysuckle, and assorted ferns and flowering plants. Spartina and rose mallow appeared in the salt marshes. As all this foliage filled these warming niches, warm-blooded animals followed, including humans.

A dearth of archeological remains suggests that the first New Yorkers probably didn't settle, but camped seasonally to pick berries, chestnuts, and wild grapes. They hunted turkey, heath hens, ducks, and white-tailed deer, but mainly they fished. The surrounding waters swarmed with smelt, shad, and herring. Brook trout ran in Manhattan streams. Oysters, clams, quahogs, crabs, and lobsters were so abundant that harvesting them was effortless. Large middens of discarded mollusk shells along the shores were the first human structures here. By the time Henry Hudson first saw the island, upper Harlem and Greenwich Village were grassy savannas, cleared repeatedly with fire by the Lenni Lenape for planting. By flooding ancient Harlem fire pits to see what floats to the top, Mannahatta Project researchers have learned that the Lenni Lenape cultivated corn, beans, squash, and sunflowers. Much of the island was still as green and dense as the Białowieża Puszcza. But well before its famous transfiguration from Indian land to colonial real estate, priced to sell at 60 Dutch guilders, the mark of *Homo sapiens* was already on Manhattan.

⤝⤞

IN THE MILLENNIAL year 2000, a harbinger of a future that might revive the past appeared in the form of a coyote that managed to reach Central Park. Subsequently, two more made it into town, as well as a wild turkey. The rewilding of New York City may not wait until people leave.

That first advance coyote scout arrived via the George Washington Bridge, which Jerry Del Tufo managed for the Port Authority of New York and New Jersey. Later, he took over the bridges that link Staten Island to the mainland and Long Island. A structural engineer in his forties, he considers

bridges among the loveliest ideas humans ever conceived, gracefully span-
ning chasms to bring people together.

Del Tufo himself spans an ocean. His olive features bespeak Sicily; his
voice is pure urban New Jersey. Bred to the pavement and steel that became his life's work, he nonetheless marvels at the annual miracle of baby
peregrine falcons hatching high atop the George Washington's towers, and
at the sheer botanical audacity of grass, weeds, and ailanthus trees that de-
fiantly bloom, far from topsoil, from metal niches suspended high above
the water. His bridges are under a constant guerrilla assault by nature. Its
arsenal and troops may seem ludicrously puny against steel-plated armor,
but to ignore endless, ubiquitous bird droppings that can snag and sprout
airborne seeds, and simultaneously dissolve paint, would be fatal. Del
Tufo is up against a primitive, but unrelenting foe whose ultimate strength
is its ability to outlast its adversary, and he accepts as a fact that ultimately
nature must win.

Not on his watch, though, if he can help it. First and foremost, he
honors the legacy he and his crew inherited: their bridges were built by a
generation of engineers who couldn't possibly have conceived of a third
of a million cars crossing them daily—yet 80 years later, they're still in
service. "Our job," he tells his men, "is to hand over these treasures to the
next generation in better shape than when we accepted them."

On a February afternoon he heads through snow flurries to the Bayonne Bridge, chatting with his crew over his radio. The underside of the
approach on the Staten Island side is a powerful steel matrix that con-
verges in a huge concrete block anchored to the bedrock, an abutment that
bears half the load of the Bayonne's main span. To stare up directly into
its labyrinthine load-bearing I-beams and bracing members, interlocked
with half-inch-thick steel plates, flanges, and several million half-inch rivets and bolts, recalls the crushing awe that humbles pilgrims gaping at the
soaring Vatican dome of St. Peter's Cathedral: something this mighty is
here forever. Yet Jerry Del Tufo knows exactly how these bridges, without
humans to defend to them, would come down.

It wouldn't happen immediately, because the most immediate threat
will disappear with us. It's not, says Del Tufo, the incessant pounding
traffic.

"These bridges are so overbuilt, traffic's like an ant on an elephant."
In the 1930s, with no computers to precisely calculate tolerances of

construction materials, cautious engineers simply heaped on excess mass and redundancy. "We're living off the overcapacity of our forefathers. The GW alone has enough galvanized steel wire in its three-inch main cables to wrap the Earth four times. Even if every other suspender rope deteriorated, the bridge wouldn't fall down."

Enemy number one is the salt that highway departments spread on the roadways each winter—ravenous stuff that keeps eating steel once it's done with the ice. Oil, antifreeze, and snowmelt dripping from cars wash salt into catch basins and crevices where maintenance crews must find and flush it. With no more people, there won't be salt. There will, however, be rust, and quite a bit of it, when no one is painting the bridges.

At first, oxidation forms a coating on steel plate, twice as thick or more as the metal itself, which slows the pace of chemical attack. For steel to completely rust through and fall apart might take centuries, but it won't be necessary to wait that long for New York's bridges to start dropping. The reason is a metallic version of the freeze-thaw drama. Rather than crack like concrete, steel expands when it warms and contracts when it cools. So that steel bridges can actually get longer in summer, they need expansion joints.

In winter, when they shrink, the space inside expansion joints opens wider, and stuff blows in. Wherever it does, there's less room for the bridge to expand when things warm up. With no one painting bridges, joints fill not only with debris but also with rust, which swells to occupy far more space than the original metal.

"Come summer," says Del Tufo, "the bridge is going to get bigger whether you like it or not. If the expansion joint is clogged, it expands toward the weakest link—like where two different materials connect." He points to where four lanes of steel meet the concrete abutment. "There, for example. The concrete could crack where the beam is bolted to the pier. Or, after a few seasons, that bolt could shear off. Eventually, the beam could walk itself right off and fall."

Every connection is vulnerable. Rust that forms between two steel plates bolted together exerts forces so extreme that either the plates bend or rivets pop, says Del Tufo. Arch bridges like the Bayonne—or the Hell Gate over the East River, made to hold railroads—are the most overbuilt of all. They might hold for the next 1,000 years, although earthquakes rippling through one of several faults under the coastal plain could shorten that period. (They would probably do better than the 14 steel-lined, concrete subway tubes beneath the East River—one of which, leading to Brooklyn, dates back to horses and buggies. Should any of their sections separate, the Atlantic Ocean would rush in.) The suspension and truss bridges that carry automobiles, however, will last only two or three centuries before their rivets and bolts fail and entire sections fall into the waiting waters.

Until then, more coyotes follow the footsteps of the intrepid ones that managed to reach Central Park. Deer, bear, and finally wolves, which have reentered New England from Canada, arrive in turn. By the time most of its bridges are gone, Manhattan's newer buildings have also been ravaged, as wherever leaks reach their embedded steel reinforcing bars, they rust, expand, and burst the concrete that sheaths them. Older stone buildings such as Grand Central—especially with no more acid rain to pock their marble—will outlast every shiny modern box.

Ruins of high-rises echo the love song of frogs breeding in Manhattan's reconstituted streams, now stocked with alewives and mussels dropped by seagulls. Herring and shad have returned to the Hudson, though they spent some generations adjusting to radioactivity trickling out of Indian Point Nuclear Power Plant, 35 miles north of Times Square, after its reinforced concrete succumbed. Missing, however, are nearly all fauna adapted to us. The seemingly invincible cockroach, a tropical import, long ago froze in unheated apartment buildings. Without garbage, rats starved or became lunch for the raptors nesting in burnt-out skyscrapers.

Rising water, tides, and salt corrosion have replaced the engineered shoreline, circling New York's five boroughs with estuaries and small beaches. With no dredging, Central Park's ponds and reservoir have been reincarnated as marshes. Without natural grazers—unless horses used by hansom cabs and by park policemen managed to go feral and breed—Central Park's grass is gone. A maturing forest is in its place, radiating down former streets and invading empty foundations. Coyotes, wolves, red foxes, and bobcats have brought squirrels back into balance with oak trees tough enough to outlast the lead we deposited, and after 500 years, even in a warming climate the oaks, beeches, and moisture-loving species such as ash dominate.

Long before, the wild predators finished off the last descendants of pet dogs, but a wily population of feral house cats persists, feeding on starlings. With bridges finally down, tunnels flooded, and Manhattan truly an island again, moose and bears swim a widened Harlem river to feast on the berries that the Lenape once picked.

Amid the rubble of Manhattan financial institutions that literally collapsed for good, a few bank vaults stand; the money within, however worthless, is mildewed but safe. Not so the artwork stored in museum vaults, built more for climate control than strength. Without electricity, protection ceases; eventually museum roofs spring leaks, usually starting with their skylights, and their basements fill with standing water. Subjected to wild swings in humidity and temperature, everything in storage rooms is prey to mold, bacteria, and the voracious larvae of a notorious museum scourge, the black carpet beetle. As they spread to other floors, fungi discolor and dissolve paintings in the Metropolitan beyond recognition. Ceramics, however, are doing fine, since they're chemically similar to fossils. Unless something falls on them first, they await reburial for the next archaeologist to dig them up. Corrosion has thickened the patina on bronze statues, but hasn't affected their shapes. "That's why we know about the Bronze Age," notes Manhattan art conservator Barbara Appelbaum.

Even if the Statue of Liberty ends up at the bottom of the harbor, Appelbaum says, its form will remain intact indefinitely, albeit somewhat chemically altered and possibly encased in barnacles. That might be the safest place for it, because at some point thousands of years hence, any stone walls still standing—maybe chunks of St. Paul's Chapel across from the World Trade Center, built in 1766 from Manhattan's own hard schist—must finally fall. Three times in the past 100,000 years, glaciers have scraped New York clean. Unless humankind's Faustian affair with carbon fuels ends up tipping the atmosphere past the point of no return, and runaway global warming transfigures Earth into Venus, at some unknown date glaciers will do so again. The mature beech-oak-ash-ailanthus forest will be mowed down. The four giant mounds of entombed garbage at the Fresh Kills landfill on Staten Island will be flattened, their vast accumulation of stubborn PVC plastic and of one of the most durable human creations of all—glass—ground to powder.

After the ice recedes, buried in the moraine and eventually in geologic layers below will be an unnatural concentration of a reddish metal, which briefly had assumed the form of wiring and plumbing. Then it was hauled to the dump and returned to the Earth. The next toolmaker to arrive or evolve on this planet might discover and use it, but by then there would be nothing to indicate that it was us who put it there.

CHAPTER 4

The World Just Before Us

1. An Interglacial Interlude

FOR MORE THAN 1 billion years, sheets of ice have been sliding back and forth from the poles, sometimes actually meeting at the equator. The reasons involve continental drift, the Earth's mildly eccentric orbit, its wobbly axis, and swings in atmospheric carbon dioxide. For the last few million years, with the continents basically where we find them today, ice ages have recurred fairly regularly and lasted upwards of 100,000 years, with intervening thaws averaging 12,000 to 28,000 years.

The last glacier left New York 11,000 years ago. Under normal conditions, the next to flatten Manhattan would be due any day now, though there's growing doubt that it will arrive on schedule. Many scientists now guess that the current intermission before the next frigid act will last a lot longer, because we've managed to postpone the inevitable by stuffing our atmospheric quilt with extra insulation. Comparisons to ancient bubbles in Antarctic ice cores reveal there's more CO_2 floating around today than at any time in the past 650,000 years. If people cease to exist tomorrow and we never send another carbon-bearing molecule skyward, what we've already set in motion must still play itself out.

That won't happen quickly by our standards, although our standards are changing, because we *Homo sapiens* didn't bother to wait until fossilization to enter geologic time. By becoming a veritable force of nature, we've already done so. Among the human-crafted artifacts that will last the longest after we're gone is our redesigned atmosphere. Thus, Tyler Volk finds nothing strange about being an architect teaching atmospheric physics and marine chemistry on the New York University biology faculty. He finds he must draw on all those disciplines to describe how humans have turned the atmosphere, biosphere, and the briny deep into something that, until now, only volcanoes and colliding continental plates have been able to achieve.

Volk is a lanky man with wavy dark hair and eyes that scrunch into crescents when he ponders. Leaning back in his chair, he studies a poster that nearly fills his office bulletin board. It portrays atmosphere and oceans as a single fluid with layers of deepening density. Until about 200 years ago, carbon dioxide from the gaseous part above dissolved into the liquid part below at a steady rate that kept the world at equilibrium. Now, with atmospheric CO_2 levels so high, the ocean needs to readjust. But because it's so big, he says, that takes time.

"Say there are no more people burning fuel. At first, the ocean's surface will absorb CO_2 rapidly. As it saturates, that slows. It loses some CO_2 to photosynthesizing organisms. Slowly, as the seas mix, it sinks, and ancient, unsaturated water rises from the depths to replace it."

It takes 1,000 years for the ocean to completely turn over, but that doesn't bring the Earth back to pre-industrial purity. Ocean and atmosphere are more in balance with each other, but both are still supercharged with CO_2. So is the land, where excess carbon will cycle through soil and life-forms that absorb but eventually release it. So where can it go? "Normally," says Volk, "the biosphere is like an upside-down glass jar: On top, it's basically closed to any extra matter, except for letting in a few meteors. At the bottom, the lid is slightly open—to volcanoes."

The problem is, by tapping the Carboniferous Formation and spewing it up into the sky, we've become a volcano that hasn't stopped erupting since the 1700s.

So next, the Earth must do what it always does when volcanoes throw extra carbon into the system. "The rock cycle kicks in. But it's much longer." Silicates such as feldspar and quartz, which comprise most of the Earth's crust, are gradually weathered by carbonic acid formed by rain and carbon dioxide, and turn to carbonates. Carbonic acid dissolves soil and minerals that release calcium to groundwater. Rivers carry this to the sea, where it precipitates out as seashells. It's a slow process, sped slightly by the intensified weather in the supercharged atmosphere.

"Eventually," Volk concludes, "the geologic cycle will take CO_2 back to prehuman levels. That will take about 100,000 years."

Or longer: One concern is that even as tiny sea creatures are locking carbon up in their armor, increased CO_2 in the oceans' upper layers may be dissolving their shells. Another is that the more oceans warm, the less CO_2 they absorb, as higher temperatures inhibit growth of CO_2-breathing plankton. Still, Volk believes, with us gone the oceans' initial 1,000-year turnover could absorb as much as 90 percent of the excess carbon dioxide, leaving the atmosphere with only about 10 to 20 extra parts per million of CO_2 above the 280 ppm preindustrial levels.

The difference between that and today's 380 ppm, scientists who've spent a decade coring the Antarctic ice assure us, means that there will be no encroaching glaciers for at least the next 15,000 years. During the time that the extra carbon is being slowly sopped up, however, palmettos and magnolias may be repopulating New York City faster than oaks and beeches. The moose may have to seek gooseberries and elderberries in Labrador, while Manhattan instead hosts the likes of armadillos and peccaries advancing from the south . . .

. . . unless, respond some equally eminent scientists who've been eying the Arctic, fresh meltwater from Greenland's ice cap chills the Gulf Stream to a halt, closing down the great ocean conveyor belt that circulates warm water around the globe. That would bring an ice age back to Europe and the East Coast of North America after all. Maybe not severe enough to trigger massive sheet glaciers, but treeless tundra and permafrost could become the alternative to temperate forest. The berry bushes would be reduced to stunted, colorful spots of ground cover among the reindeer lichen, attracting caribou southward.

In a third, wishful scenario, the two extremes might blunt each other enough to hold temperatures suspended in between. Whichever it is, hot or cold or betwixt, in a world where humans stayed around and pushed atmospheric carbon to 500 or 600 parts per million—or the projected 900 ppm by AD 2100, if we change nothing from the way we do business today—much of what once lay frozen atop Greenland will be sloshing in a swollen Atlantic. Depending on exactly how much of it and Antarctica go, Manhattan might become no more than a few islets, one where the Great Hill once rose above Central Park, another an outcropping of schist in Washington Heights. For a while, clutches of buildings a few miles to the south would vainly scan the surrounding waters like surfacing periscopes, until buffeting waves brought them down.

2. Ice Eden

Had humans never evolved, how might the planet have fared? Or was it inevitable that we did?

And if we disappeared, would—or could—we, or something equally complicated, happen again?

❧

FAR FROM EITHER pole, East Africa's Lake Tanganyika lies in a crack that, 15 million years ago, began to split Africa in two. The Great African Rift Valley is the continuation of a tectonic parting of the ways that began even earlier in what is now Lebanon's Beqaa Valley, then ran south to form the course of the Jordan River and the Dead Sea. Then it widened into the Red Sea, and is now branching down two parallel cleavages through the crust of Africa. Lake Tanganyika fills the Rift's western fork for 420 miles, making it the longest lake in the world.

Nearly a mile from surface to bottom, around 10 million years old, it is also the world's second-deepest and second-oldest, after Siberia's Lake Baikal. That makes it extremely interesting to scientists who have been extracting core samples of the lake bed sediments. Just as annual snowfalls preserve a history of climate in glaciers, pollen grains from surrounding foliage settle in the depths of bodies of freshwater, neatly separated into readable layers by dark bands of rainy-season runoff and light seams of dry-season algal blooms. At ancient Lake Tanganyika, the cores reveal more than the identities of plants. They show how a jungle gradually turned to fire-tolerant, broad-leafed woodland known as miombo, which covers vast swathes of today's Africa. Miombo is another man-made artifact, which developed as paleolithic humans discovered that by burning trees they could create grassland and open woodlands to attract and nurture antelope.

Mixed with thickening layers of charcoal, the pollens show the even greater deforestation that accompanied the dawning of the Iron Age, as humans learned first to smelt ore, then to fashion hoes for furrows. There they planted crops such as finger millet, whose signature also appears. Later arrivals, like beans and corn, produce either too few pollens or grains too large to drift far, but the spread of agriculture is evidenced by the increase of pollens from ferns that colonize disturbed land.

All this and more can be learned from mud recovered with 10 meters of steel pipe lowered on a cable and, aided by a vibrating motor, driven by the force of its own weight into the lake bed—and into 100,000 years of pollen layers. A next step, says University of Arizona paleolimnologist Andy Cohen, who heads a research project in Kigoma, Tanzania, on Lake Tanganyika's eastern shore, is a drill rig capable of penetrating a 5-million- or even 10-million-year core.

Such a machine would be very expensive, on the order of a small oil-drilling barge. The lake is so deep that the drill could not be anchored, requiring thrusters linked to a global positioning system to constantly adjust its position above the hole. But it would be worth it, says Cohen, because this is Earth's longest, richest climate archive.

"It's long been assumed that climate is driven by advancing and retreating polar ice sheets. But there's good reason to believe that circulation at the tropics is also involved. We know a lot about climate change at the poles, but not at the heat engine of the planet, where people live." Coring it, Cohen says, would capture "ten times the climate history found in glaciers, and with far greater precision. There are probably a hundred different things we can analyze."

Among them is the history of human evolution, because the core's record would span the years during which primates took their first bipedal steps and proceeded through transcendent stages that led to hominids from *Australopithecus* to *Homos habilis, erectus,* and finally *sapiens*. The pollens would be the same that our ancestors inhaled, even broadcast from the same plants they touched and ate, because they, too, emerged from this Rift.

East of Lake Tanganyika in the African Rift's parallel branch, another lake, shallower and saline, evaporated and reappeared various times over the past 2 million years. Today, it is grassland, hard-grazed by the cows and goats of Maasai herdsmen, overlying sandstone, clay, tuff, and ash atop a bed of volcanic basalt. A stream draining Tanzania's volcanic highlands to the east gradually cut a gorge through those layers 100 meters deep. There, during the 20th century, archaeologists Louis and Mary Leakey discovered fossilized hominid skulls left 1.75 million years earlier. The gray rubble of Olduvai Gorge, now a semidesert bristling with sisal, eventually yielded hundreds of stone-flake tools and chopper cores made from the underlying basalt. Some of these have been dated to 2 million years ago.

In 1978, 25 miles southwest of Olduvai Gorge, Mary Leakey's team found a trail of footprints frozen in wet ash. They were made by an australopithecine trio, likely parents and a child, walking or fleeing through the rainy aftermath of an eruption of the nearby Sadiman volcano. Their discovery pushed bipedal hominid existence back beyond 3.5 million years ago. From here and from related sites in Kenya and Ethiopia, a pattern emerges of the gestation of the human race. It is now known that we walked on two feet for hundreds of thousands of years before it occurred to us to strike one stone against another to create sharp-edged tools. From the remains of hominid teeth and other nearby fossils, we know we were omnivores, equipped with molars to crunch nuts—but also, as we advanced from finding stones shaped like axes to learning how to produce them, possessed of the means to efficiently kill and eat animals.

Olduvai Gorge and the other fossil hominid sites, together comprising a crescent that runs south from Ethiopia and parallels the continent's eastern shore, have confirmed beyond much doubt that we are all Africans. The dust we breathe here, blown by zephyrs that leave a coating of gray tuff powder on Olduvai's sisals and acacias, contains calcified specks of the very DNA that we carry. From this place, humans radiated across continents and around a planet. Eventually, coming full circle, we returned, so estranged from our origins that we enslaved blood cousins who stayed behind to maintain our birthright.

Animal bones in these places—some from hippo, rhino, horse, and elephant species that became extinct as we multiplied; many of them honed by our ancestors into pointed tools and weapons—help us know how the world was just before we emerged from the rest of *Mammalia*. What they don't show, however, is what might have impelled us to do so. But at Lake Tanganyika, there are some clues. They lead back to the ice.

The lake is fed by many streams that pour off the mile-high Rift escarpment. At one time, these dropped through gallery rain forest. Then came miombo woodland. Today, most of the escarpment has no trees at all. Its slopes have been cleared to plant cassava, with fields so steep that farmers are known to roll off them.

An exception is at Gombe Stream, on Lake Tanganyika's eastern Tanzanian coast, the site where primatologist Jane Goodall, a Leakey assistant at Olduvai Gorge, has studied chimpanzees since 1960. Her field study, the longest anywhere of how a species behaves in the wild, is headquartered in a camp reachable only by boat. The national park that surrounds it is Tanzania's smallest—only 52 square miles. When Goodall first arrived, the surrounding hills were covered in jungle. Where it opened into woodland and savanna, lions and cape buffalo lived. Today, the park is surrounded on three sides by cassava fields, oil palm plantations, hill settlements, and, up and down the lakeshore, several villages of more than 5,000 inhabitants. The famous chimpanzee population teeters precariously around 90.

Although chimps are the most intensely studied primates at Gombe, its rain forest is also home to many olive baboons and several monkey species: vervet, red colobus, red-tailed, and blue. During 2005, a Ph.D. candidate at New York University's Center for the Study of Human Origins named Kate Detwiler spent several months investigating an odd phenomenon involving the last two.

Red-tailed monkeys have small black faces, white-spotted noses, white cheeks, and vivid chestnut tails. Blue monkeys have bluish coats and triangular, nearly naked faces, with impressive jutting eyebrows. With different coloring, body size and vocalizations, no one would confuse blue and red-tailed monkeys in the field. Yet in Gombe they now apparently mistake one another, because they have begun to interbreed. So far, Detwiler has confirmed that although the two species have different numbers of chromosomes, at least some of the offspring of these liaisons—whether between blue males and red-tailed females or vice versa—are fertile. From the forest floor, she scrapes their feces, in which fragments of intestinal lining attest to a mix of DNA resulting in a new hybrid.

Only she thinks it's something more. Genetics indicate that at some point 3 million to 5 million years ago, two populations of a species that was the common ancestor to these two monkeys became separated. Adjusting to distinct environments, they gradually diverged from each other. Through a similar situation involving finch populations that became isolated on various Galápagos islands, Charles Darwin first deduced how evolution works. In that case, 13 different finch species emerged in response to locally available food, their bills variously adapted to cracking seeds, eating insects, extracting cactus pulp, or even sucking the blood of seabirds.

In Gombe, the opposite has apparently occurred. At some point, as new forest filled the barrier that once divided these two species, they found themselves sharing a niche. But then they became marooned together, as the forest surrounding Gombe National Park gave way to cassava croplands. "As the number of available mates of their own species dwindled," Detwiler figures, "these animals have been driven to desperate—or creative—survival measures."

Her thesis is that hybridization between two species can be an evolutionary force, just like natural selection is within one. "Maybe at first the mixed offspring isn't as fit as either parent," she says. "But for whatever reason—constrained habitat, or low numbers—the experiment keeps getting repeated, until eventually a hybrid as viable as its parent emerges. Or, maybe even with advantage over the parents, because the habitat has changed."

That would make the future offspring of these monkeys human artifacts: their parents forced together by agricultural *Homo sapiens* who so fragmented east Africa that populations of monkeys and other species like shrikes or flycatchers had to interbreed, crossbreed, perish—or do something very creative. Such as evolve.

Something similar may have happened here before. Once, when its Rift was only beginning to form, Africa's tropical forest filled the continent's midriff from the Indian Ocean to the Atlantic. Great apes had already made their appearance, including one that in many ways resembled chimpanzees. No remnants of it have ever been found, for the same reason that chimp remains are so rare: in tropical forests, heavy rains leach minerals from the ground before anything can fossilize, and bones decompose quickly. Yet scientists know it existed, because genetics show that we and chimps descended directly from the same ancestor. The American physical anthropologist Richard Wrangham has given this undiscovered ape a name: *Pan prior.*

Prior, that is, to *Pan troglodytes,* today's chimpanzee, but also prior to a great dry spell that overtook Africa about 7 million years ago. Wetlands retreated, soils dried, lakes disappeared, and forests shrank into pocket refuges, separated by savannas. What caused this was an ice age advancing from the poles. With much of the world's moisture locked into glaciers that buried Greenland, Scandinavia, Russia, and much of North America, Africa became parched. No ice sheets reached it, although glacial caps formed on volcanoes like Kilimanjaro and Mount Kenya. But the climatic change that fragmented Africa's forest, more than twice the size of today's Amazon, was due to the same distant white juggernaut that was smashing conifers in its path.

That faraway ice sheet stranded populations of African mammals and birds in patches of forest where, over the next few million years, they evolved their separate ways. At least one of them, we know, was driven to try something daring: taking a stroll in a savanna.

If humans vanished, and if something eventually replaced us, would it begin as we did? In southwest Uganda, there's a place where it's possible to see our history reenacted in microcosm. Chambura Gorge is a narrow ravine that cuts for 10 miles through a deposit of dark brown volcanic ash on the floor of the Rift Valley. In startling contrast to the surrounding yellow plains, a green band of tropical sobu, ironwood, and leafflower trees fills this canyon along the Chambura River. For chimpanzees, this oasis is both a refuge and a crucible. Lush as it is, the gorge is barely 500 yards across, its available fruit too limited to satisfy all their dietary needs. So from time to time, brave ones risk climbing up the canopy and leaping to the rim, to the chancy realm of the ground.

With no ladder of branches to help them see over the oat and citronella grasses, they must raise themselves on two feet. Perched for a moment on the verge of being bipedal, they scan for lions and hyenas among the scattered fig trees on the savanna. They select a tree they calculate they can reach without becoming food themselves. Then, as we also once did, they run for it.

About 3 million years after distant glaciers pushed some courageous, hungry specimens of *Pan prior* out of forests no longer big enough to sustain us—and some of them proved imaginative enough to survive—the world warmed again. Ice retreated. Trees regained their former ground and then some, even covering Iceland. Forests reunited across Africa, again from the Atlantic shore to the Indian, but by then *Pan prior* had segued into something new: the first ape to prefer the grassy woodlands at the forest's edge. After more than a million years of walking on two feet, its legs had lengthened and its opposable big toes had shortened. It was losing the ability to dwell in trees, but its sharpened survival skills on the ground had taught it to do so much more.

Now we were hominids. Somewhere along the way, as *Australopithecus* was begetting *Homo,* we learned not only to follow the fires that opened up savannas that we'd learned to inhabit, but how to make them ourselves. For some 3 million years more, we were too few to create more than local patchworks of grassland and forests whenever distant ice ages weren't doing it for us. Yet in that time, long before *Pan prior*'s most recent descendant, surnamed *sapiens,* appeared, we must have become numerous enough to again try being pioneers.

Were the hominids who wandered out of Africa again intrepid risktakers, their imaginations picturing even more bounty beyond the savanna's horizon?

Or were they losers, temporarily out-competed by tribes of stronger blood cousins for the right to stay in our cradle?

Or were they simply going forth and multiplying, like any beast presented with rich resources, such as grasslands stretching all the way to Asia? As Darwin came to appreciate, it didn't matter: when isolated groups from the same species proceed in their separate ways, the most successful among them learn to flourish in new surroundings. Exiles or adventurers, the ones who survived filled Asia Minor and then India. In Europe and Asia, they began to develop a skill long known to temperate creatures like squirrels but new to primates: *planning,* which required both memory and foresight to store food in seasons of plenty in order to outlast seasons of cold. A land bridge got them through much of Indonesia, but to reach New

Australopithecus africanus.

ILLUSTRATION BY CARL BUELL.

Guinea and, about 50,000 years ago, Australia, they had to learn to become seafarers. And then, 11,000 years ago, observant *Homo sapiens* in the Middle East figured out a secret until then known only to select species of insects: how to control food supplies not by destroying plants, but by nurturing them.

Because we know the Middle Eastern origin of the wheat and barley they grew, which soon spread southward along the Nile, we can guess that—like shrewd Jacob returning with a cornucopia of gifts to win over his powerful brother, Esau—someone bearing seeds and the knowledge of agriculture returned from there to the African homeland. It was an auspicious time to do so, because yet another ice age—the last one—had once again stolen moisture from lands that glaciers didn't reach, tightening food supplies. So much water was frozen into glaciers that the oceans were 300 feet lower than today.

At that same time, other humans who had kept spreading across Asia arrived at the farthest reach of Siberia. With the Bering Sea partly emptied, a land bridge 1,000 miles across connected to Alaska. For 10,000 years, it had lain under more than half a mile of ice. But now, enough had receded to reveal an ice-free corridor, in places 30 miles wide. Picking their way around meltwater lakes, they crossed it.

Chambura Gorge and Gombe Stream are atolls in an archipelago that is all that remains of the forest that birthed us. This time, the fragmentation of Africa's ecosystem is due not to glaciers, but to ourselves, in our latest evolutionary leap to the status of Force of Nature, having become as powerful as volcanoes and ice sheets. In these forest islands, surrounded by seas of agriculture and settlement, the last of *Pan prior*'s other offspring still cling to life as it was when we left to become woodland, savanna, and finally city apes. To the north of the Congo River, our siblings are gorillas and chimpanzees; to the south, bonobos. It is the latter two we genetically most resemble; when Louis Leakey sent Jane Goodall to Gombe, it was because the bones and skulls he and his wife had uncovered suggested that our common ancestor would have looked and acted much like chimpanzees.

Whatever inspired our forebears to leave, their decision ignited an evolutionary burst unlike any before, described variously as the most successful and most destructive the world has ever seen. But suppose we had stayed—or suppose that, when we were exposed on the savanna, the ancestors of today's lions and hyenas had made short work of us. What, if anything, would have evolved in our place?

To stare into the eyes of a chimpanzee in the wild is to glimpse the world had we stayed in the forest. Their thoughts may be obscure, but their intelligence is unmistakable. A chimpanzee in his element, regarding you coolly from a branch of an mbula fruit tree, expresses no sense of inferiority in the presence of a superior primate. Hollywood images mislead, because its trained chimps are all juveniles, as cute as any child. However, they keep growing, sometimes reaching 120 pounds. In a human of similar weight, about 30 of those pounds would be fat. A wild chimpanzee, who lives in a perpetual state of gymnastics, has perhaps three to four pounds of fat. The rest is muscle.

Dr. Michael Wilson, the curly-haired young director of field research at Gombe Stream, vouches for their strength. He has watched them tear apart and devour red colobus monkeys. Superb hunters, about 80 percent of their attempts are successful kills. "For lions, it's only about one out of 10 or 20. These are pretty bright creatures."

But he has also seen them steal into the territories of neighboring chimp groups, ambush unwary lone males, and maul them to death. He's watched chimps over months patiently pick off males of neighboring clans until the territory and the females are theirs. He's also seen pitched chimpanzee combat, and blood battles within a group to determine who is the alpha male. The unavoidable comparisons to human aggression and power struggles became his research specialty.

"I get tired of thinking about it. It's kind of depressing."

One of the unfathomables is why bonobos, smaller and more slender than chimps but equally related to us, don't seem very aggressive at all. Although they defend territory, no intergroup killing has ever been observed. Their peaceful nature, predilection for playful sex with multiple partners, and apparent matriarchal social organization with all the attendant nurturing have practically become mythologized among those who insistently hope that the meek might yet inherit the Earth.

In a world without humans, however, if they had to fight it out with chimpanzees, they would be outnumbered: only 10,000 or fewer bonobos remain, compared to 150,000 chimpanzees. Since their combined population a century ago was approximately 20 times greater, with each passing year chances weaken for either species to be around to take over.

Michael Wilson, hiking in the rain forest, hears drumbeats that he knows are chimpanzees pounding on buttress roots, signaling each other. He runs after them, up and down Gombe's 13 stream valleys, hurdling morning glory vines and lianas strung across baboon trails, following chimp hoots until, two hours later, he finally catches them at the top of the Rift. Five of them are in a tree at the edge of the woodland, eating the mangoes they love, a fruit that came along with wheat from Arabia.

A mile below, Lake Tanganyika flashes in the afternoon sunlight, so vast it holds 20 percent of the world's freshwater and so many endemic fish species that it's known among aquatic biologists as the Galápagos of lakes. Beyond it, to the west, are the hazy hills of the Congo, where chimpanzees are still taken for bush meat. In the opposite direction, past Gombe's boundaries, are farmers who also have rifles, and who are tired of chimps who snatch their oil palm nuts.

Other than humans and each other, the chimpanzees have no real predators here. The very presence of these five in a tree surrounded by grass testifies to the fact that they've also inherited the gene of adaptability, and are far more able than gorillas, which have highly specialized forest diets, to live on a variety of foods and in a variety of environments. If humans were gone, however, they might not need to. Because, says Wilson, the forest would come back. Fast.

"There'd be miombo moving through the area, recovering the cassava fields. Probably the baboons would take first advantage, radiating out, carrying seeds in their poop, which they'd plant. Soon you'd have trees sprouting wherever there's suitable habitat. Eventually, the chimps would follow."

With plenty of game returning, lions would find their way back, then the big animals: cape buffalo and elephant, coming from Tanzania's and Uganda's reserves. "Eventually," Wilson says, sighing, "I can see a continuous stretch of chimpanzee populations, all the way down to Malawi, all the way up to Burundi, and over into Congo."

All that forest back again, ripe with chimpanzees' favorite fruits and a

prospering population of red colobus to catch. In tiny Gombe—a protected shred of Africa's past that is also a taste of such a posthuman
future—no enticement is readily evident for another primate to leave all
that lushness and follow in our futile footsteps.

Until, of course, the ice returns.

CHAPTER 5

The Lost Menagerie

IN A DREAM, you walk outside to find your familiar landscape swarming with fantastic beings. Depending on where you live, there might be deer with antlers thick as tree boughs, or something resembling a live armored tank. There's a herd of what look like camels—except they have trunks. Furry rhinoceroses, big hairy elephants, and even bigger sloths—sloths?? Wild horses of all sizes and stripe. Panthers with seven-inch fangs and alarmingly tall cheetahs. Wolves, bears, and lions so huge, this must be a nightmare.

A dream, or a congenital memory? This was precisely the world that *Homo sapiens* stepped into as we spread beyond Africa, all the way to America. Had we never appeared, would those now-missing mammals still be here? If we go, will they be back?

⤜∘⤛

AMONG THE VARIOUS slurs hurled at sitting presidents throughout the history of the United States, the epithet with which Thomas Jefferson's foes smeared him in 1808 was unique: "Mr. Mammoth." An embargo that Jefferson slapped on all foreign trade, intended to punish Britain and France for monopolizing shipping lanes, had backfired. While the U.S. economy was collapsing, his opponents sneered, President Jefferson could be found in the East Room of the White House, playing with his fossil collection.

This was true. Jefferson, a passionate naturalist, had been enthralled for years by reports of huge bones strewn around a salt lick in the Kentucky wilderness. Descriptions suggested that they were similar to remains discovered in Siberia of a species of giant elephant, thought by European scientists to be extinct. African slaves had recognized big molars found in the Carolinas as belonging to some kind of elephant, and Jefferson was sure these were the same. In 1796 he received a shipment, supposedly of mammoth bones, from Greenbriar County, Virginia, but a huge claw immediately alerted him that this was something else, possibly some immense breed of lion. Consulting anatomists, he eventually identified it and is credited for the first description of a North America ground sloth, today named *Megalonyx jefersoni.*

Most exciting to him, though, were testimonies by Indians near the Kentucky salt lick, allegedly corroborated by other tribes farther west, that the tusked behemoth in question still lived in the north. After he became president, he sent Meriwether Lewis to study the Kentucky site on the way to joining William Clark for their historic mission. Jefferson had charged Lewis and Clark not only with traversing the Louisiana Purchase and seeking a northwest river route to the Pacific, but also with finding live mammoths, mastodons, or anything similarly large and unusual.

That part of their otherwise stunning expedition proved a failure; the most impressive big mammal they cited was the bighorn sheep. Jefferson later contented himself with sending Clark back to Kentucky for the mammoth bones that he displayed in the White House, today part of museum collections in the United States and France. He is often credited with founding the science of paleontology, though it was not really his intention. He'd hoped to belie an opinion, espoused by a prominent French scientist, that everything in the New World was inferior to the Old, including its wildlife.

He was also fundamentally mistaken about the meaning of fossil bones: he was convinced that they must belong to a living species, because he didn't believe that anything ever went extinct. Although often considered America's quintessential Age of Enlightenment intellectual, Jefferson's beliefs corresponded to those held by many Deists and Christians of his day: that in a perfect Creation, nothing created was ever intended to disappear.

He articulated this credo, however, as a naturalist: "Such is the economy of nature that no instance can be produced of her having permitted any one race of her animals to become extinct." It was a wish that imbued many of his writings: he wanted these animals to be alive, wanted to know them. His quest for knowledge led him to found the University of Virginia. Over the next two centuries, paleontologists there and elsewhere would show that many species had in fact died. Charles Darwin would describe how these extinctions were part of nature itself—one variety morphing into the next to meet changing conditions, another losing its niche to a more powerful competitor.

Yet one detail that nagged at Thomas Jefferson and others after him was that the big-mammal remains turning up didn't seem all that old. These weren't heavily mineralized fossils embedded in solid layers of rock. Tusks, teeth, and jawbones at places like Kentucky's Big Bone Lick were still strewn on the ground, or protruding from shallow silt, or on the floor of caves. The big mammals they belonged to couldn't have been gone that long. What had happened to them?

The Desert Laboratory—originally the Carnegie Desert Botanical Laboratory—was built more than a century ago on Tumamoc Hill, a butte in southern Arizona overlooking what was then one of North America's finest stands of cactus forest and, beyond that, Tucson. For nearly half the lab's existence, a tall, broad-shouldered, affable paleoecologist named Paul Martin has been here. During that time the desert below Tumamoc's saguaro-covered slopes disappeared under a snarl of dwellings and commerce. Today, the Lab's fine old stone structures occupy what developers today consider prime view property, which they continually scheme to wrest from its present owner, the University of Arizona. Yet when Paul Martin leans on his cane to gaze out his lab's screened doorway, his frame of reference for human impact is not merely the past century, but the last 13,000 years—since people came to stay.

In 1956, a year before arriving here, Paul Martin had spent the winter in a Quebec farmhouse, during a postdoctoral fellowship at the University of Montreal. A case of polio contracted while collecting bird specimens in Mexico as a zoology undergraduate had rerouted his research from the field to the laboratory. Holed up in Canada with a microscope, he studied sediment cores from New England lakes that dated back to the end of the last ice age. The samples revealed how, as the climate softened, surrounding vegetation changed from treeless tundra to conifers to temperate deciduous—a progression some suspected led to mastodon extinction.

One snowbound weekend, weary of counting tiny grains of pollen, he opened a taxonomy text and started tallying the number of mammals that had disappeared in North America over the past 65 million years. When he reached the final three millennia of the Pleistocene epoch, which lasted from 1.8 million until 10,000 years ago, he started to notice something odd.

During the time frame that coincided with his sediment samples, starting about 13,000 years ago, an explosion of extinctions had occurred. By the beginning of the next epoch—the Holocene, which continues today—nearly 40 species had disappeared, all of them large terrestrial mammals. Mice, rats, shrews, and other small fur-bearing creatures had emerged unscathed, as had marine mammals. Terrestrial megafauna, however, had taken an enormous, lethal wallop.

Among the missing were a legion of animal kingdom Goliaths: giant armadillos and the even-bigger glyptodonts, resembling armor-plated Volkswagens, with tails that ended in spiked maces. There were giant short-faced bears, nearly double the size of grizzlies and, with extra long limbs, much faster—one theory suggests that giant short-faced bears in Alaska were why Siberian humans hadn't crossed the Bering Strait much earlier. Giant beavers, as big as today's black bears. Giant peccaries, which may have been prey to *Panthera leo atrox,* the American lion that was considerably bigger and swifter than today's surviving African species. Likewise, the dire wolf, the largest of canines, with a massive set of fangs.

The best-known extinct colossus, the northern woolly mammoth, was only one of many kinds of *Proboscidea,* including the imperial mammoth, largest of all at 10 tons; the hairless Columbian mammoth, which lived in warmer regions; and, in California's Channel Islands, a dwarf mammoth no taller than a human—only the collie-sized elephants on Mediterranean islands were smaller. Mammoths were grazing animals, evolved to steppes, grassland, and tundra, unlike their much older relatives, the mastodons, which browsed in woods and forests. Mastodons had been around for 30 million years, and ranged from Mexico to Alaska to Florida—but suddenly they, too, were gone. Three genera of American horses: gone. Multiple varieties of North American camels, tapirs, numerous antlered creatures ranging from dainty pronghorns to the stag moose, which resembled a cross between a moose and an elk but was larger than either, all gone, along with the saber-toothed tiger and the American cheetah (the reason why the sole remaining pronghorn species of antelope is so fleet). All gone. And all pretty much at once. What, Paul Martin wondered, could possibly have caused that?

The following year, he was on Tumamoc Hill, his big frame again perched over a microscope. This time, rather than pollen grains saved from decay by an airtight covering of lake-bottom silt, he was viewing magnified fragments preserved in a moisture-free Grand Canyon cave. Soon after he arrived in Tucson, his new boss at the Desert Lab had handed him an earthen gray lump the approximate size and shape of a softball. It was at least 10,000 years old, but unmistakably a turd. Mummified but not mineralized, it yielded identifiable fibers of grasses and flowering globe mallow. The plentiful juniper pollen Martin found confirmed his subject's great age: temperatures near the floor of the Grand Canyon have not been cool enough to sustain juniper for eight millennia.

The beast that excreted it was a Shasta ground sloth. Today, the only surviving sloths are two tree-dwelling species found in the Central and South American tropics, small and light enough to quietly inhabit rain forest canopies far from the ground, out of harm's way. This one, however, was the size of a cow. It walked on its knuckles like another of its surviving relatives, the giant South American anteater, to protect the claws it used to forage and to defend itself. It weighed half a ton, yet it was the smallest of the five sloth species that lumbered around North America, from the Yukon to Florida. The Florida variety, the size of a modern elephant, topped three tons. That was only half the size of a ground sloth in Argentina and Uruguay, which at 13,000 pounds stood taller than the largest mammoth.

A decade would pass before Paul Martin got to visit the opening in the red Grand Canyon sandstone wall above the Colorado River where his first sloth dung ball had been collected. By then, extinct American ground sloths had come to mean much more to him than simply more oversized mammals that had mysteriously toppled into oblivion. The fate of sloths would provide what Martin believed was conclusive proof of a theory forming in his mind as data accumulated like layers of stratified sediment. Inside Rampart Cave was a mound of dung deposited, he and his colleagues concluded, by untold generations of female sloths who took shelter there to give birth. The manure pile was five feet high, 10 feet across, and more than 100 feet long. Martin felt like he'd entered a sacred place.

When vandals set it on fire 10 years later, the fossil dung heap was so enormous that it burned for months. Martin mourned, but by then he had been setting blazes of his own in the paleontology world with his theory of what had wiped out millions of ground sloths, wild pigs, camels, *Proboscidea*, multiple species of horses—at least 70 entire genera of large mammals throughout the New World, all vanished in a geologic twinkling of about 1,000 years:

"It's pretty simple. When people got out of Africa and Asia and reached other parts of the world, all hell broke loose."

Martin's theory, soon dubbed the Blitzkrieg by its supporters and detractors alike, contended that, starting with Australia about 48,000 years ago, as humans arrived on each new continent they encountered animals that had no reason to suspect that this runty biped was particularly threatening. Too late, they learned otherwise. Even when hominids were still *Homo erectus*, they had already been mass-producing axes and cleavers in Stone Age factories, such as the one at Olorgesailie, Kenya, discovered a million years later by Mary Leakey. By the time a group of them arrived at the threshold of America 13,000 years ago, they had been *Homo sapiens* for at least 50,000 years. Using their bigger brains, humans by then had mastered not just the technology of attaching fluted stone points to wooden shafts, but also the atlatl, a handheld wooden lever that enabled them to propel a spear fast and precisely enough to fell dangerously large animals from a relatively safe distance.

The first Americans, Martin believes, were the ones who expertly produced the leaf-shaped flint projectile points found widely throughout North America. Both the people and their lithic points are known as Clovis, named for the New Mexico site where they were first discovered. Radiocarbon dates of organic matter found in Clovis sites have sharpened past estimates, and archaeologists now agree that Clovis people were in America 13,325 years ago. What exactly their presence signifies is, however, still a matter for hot dispute, beginning with Paul Martin's premise that humans perpetrated the extinctions that killed off three-fourths of

Litoptern.
Macrauchenia patachonica.
ILLUSTRATION BY CARL BUELL.

America's late Pleistocene megafauna, a menagerie far richer than Africa's today.

Key to Martin's Blitzkrieg theory is that in at least 14 of those sites, Clovis points were found with mammoth or mastodon skeletons, some stuck between their ribs. "If *Homo sapiens* had never evolved," he says, "North America would have three times as many animals over one ton as Africa today." He ticks off Africa's current five: "Hippos, elephants, giraffes, two rhinoceroses. We'd have 15. Even more, when we add South America. There were amazing mammals down there. Litopterns that looked like a camel with nostrils on top of their nose rather than on the tip. Or toxodons, one-ton brutes like a cross between a rhino and hippo, but anatomically neither."

All these existed, the fossil record shows, but not everyone agrees on what happened to them. One challenge to Paul Martin's theory questions whether Clovis people were actually the first humans to enter the New World. Among the objectors are Native Americans wary of any suggestion that they immigrated, which would undermine their indigenous status; they denounce the idea that their origins trace to a Bering land bridge as an attack on their faith. Even some archaeologists question whether a

Bering ice-free corridor really existed, and suggest that the first Americans actually arrived by water, skirting the ice sheet to continue down the Pacific coast. If boats reached Australia from Asia nearly 40 millennia earlier, why not boats between Asia and America?

Still others point to a handful of archaeological sites that supposedly predate Clovis. Archaeologists who excavated the most famous of these, Monte Verde, in southern Chile, believe that humans may have settled there twice: once 1,000 years prior to Clovis, the other time 30,000 years ago. If so, at that time the Bering Strait would likely not have been dry land, meaning an ocean voyage from some direction was involved. Even the Atlantic has been suggested, by archaeologists who think that Clovis techniques for flaking chert resemble paleolithics that developed in France and Spain 10,000 years earlier.

Questions about the validity of Monte Verde's radiocarbon dates soon cast doubt over initial claims that it proved early human presence in the Americas. Matters were further muddied when most of the peat bog that had preserved Monte Verde's poles, stakes, spear points, and knotted grasses was bulldozed before other archaeologists could examine the excavation site.

Even if early humans did somehow find their way to Chile before Clovis, argues Paul Martin, their impact was brief, local, and ecologically negligible, like that of the Vikings who colonized Newfoundland before Columbus. "Where are the abundant tools, artifacts, and cave paintings that their contemporaries left all over Europe? Pre-Clovis Americans wouldn't have met competing human cultures, like the Vikings did. Only animals. So why didn't they spread?"

The second, more fundamental controversy about Martin's Blitzkrieg theory, for years the most accepted explanation for the fate the of the New World's big animals, asks how a few nomadic bands of hunter-gatherers could annihilate tens of millions of large animals. Fourteen kill sites on an entire continent hardly add up to megafaunal genocide.

Nearly half a century later, the debate Paul Martin ignited remains one of science's greatest flash points. Careers have been built upon proving or attacking his conclusions, fueling a protracted, not-always-polite war waged by archaeologists, geologists, paleontologists, dendro- and radiochronologists, paleoecologists, and biologists. Nevertheless, nearly all are Martin's friends, and many are his former students.

The leading alternatives they've proposed to his overkill theory involve either climate change or disease, and have inevitably come to be known as "over-chill" and "over-ill." Over-chill, with the greatest number of adherents, is partly a misnomer, because both overheating and overcooling get blamed. In one argument, a sudden temperature reversal at the end of the Pleistocene, just as glaciers were melting away, plunged the world briefly back into the Ice Age and caught millions of vulnerable animals unaware. Others propose the opposite: that rising Holocene temperatures doomed furry species, because they had adapted over thousands of years to frigid conditions.

Over-ill suggests that arriving humans, or creatures that accompanied them, introduced pathogens that nothing alive in the Americas had ever encountered. It may be possible to prove this by analyzing mammoth tissues that will likely be discovered as glaciers continue to thaw. The premise has a grim analog: Most descendants of whoever were the first Americans died horribly in the century following European contact. Only a tiny fraction lost their lives to the point of a Spanish sword; the rest succumbed to Old World germs for which they had no antibodies: smallpox, measles, typhoid, and whooping cough. In Mexico alone, where an estimated 25 million Meso-Americans lived when the Spaniards first appeared, only 1 million remained 100 years later.

Even if disease mutated from humans to mammoths and the other Pleistocene giants, or passed directly from their dogs or livestock, that would still put the blame on *Homo sapiens*. As for over-chill, Paul Martin replies: "To quote some paleo-climate experts, 'Climate change is redundant.' It's not that the climate doesn't change, but that it changes so often."

Ancient European sites show that *Homo sapiens* and *Homo neanderthalensis* both drifted north or south with advancing or receding ice sheets. Megafauna, Martin says, would have done the same. "Large animals are buffered against temperature by their size. And they can migrate long distances—maybe not as far as birds, but compared to a mouse, pretty well. Since mice, pack rats, and other small, warm-blooded creatures survived the Pleistocene extinctions," he adds, "it's hard to believe that a sudden climate shift made life intolerable for big mammals."

Plants, even less mobile than animals, and generally more climate-sensitive, also seem to have survived. Among the sloth dung in Rampart and other Grand Canyon caves, Martin and his colleagues encountered ancient pack rat middens layered with thousands of years of vegetation remains. With the possible exception of a single variety of spruce, no species harvested by pack rat or sloth residents of these caves met temperatures extreme enough to spell their extinction.

But the clincher for Martin is the sloths. Within a millennium of the Clovis people's appearance, every slow, plodding, easy target of a ground sloth was gone—on the continents of North and South America. Yet radiocarbon dates confirm that bones found in caves in Cuba, Haiti, and Puerto Rico belonged to ground sloths still alive 5,000 years later. Their ultimate disappearance coincided with the eventual arrival of humans in the Greater Antilles 8,000 years ago. In the Lesser Antilles, on islands that humans reached even later, like Grenada, the sloth remains are even younger.

"If a change in climate was powerful enough to exterminate ground sloths from Alaska to Patagonia, you'd expect it would also take them out in the West Indies. But that didn't happen." This evidence also suggests that the first Americans arrived on the continent on foot, not as seafarers, since it took them five millennia to reach the Caribbean.

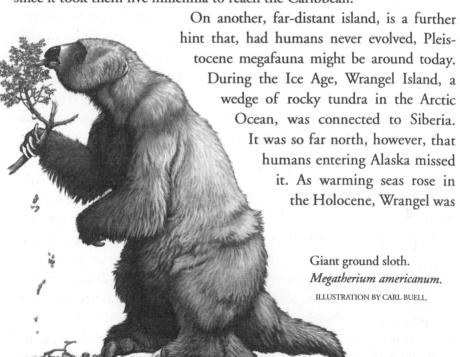

On another, far-distant island, is a further hint that, had humans never evolved, Pleistocene megafauna might be around today. During the Ice Age, Wrangel Island, a wedge of rocky tundra in the Arctic Ocean, was connected to Siberia. It was so far north, however, that humans entering Alaska missed it. As warming seas rose in the Holocene, Wrangel was

Giant ground sloth.
Megatherium americanum.
ILLUSTRATION BY CARL BUELL.

again isolated from the mainland; its population of woolly mammoths, spared but now stranded, was forced to adapt to the limited resources of an island. During the span in which humans went from caves to building great civilizations in Sumer and Peru, Wrangel Island's mammoths lived on, a dwarf species that lasted 7,000 years longer than mammoths on any continent. They were still alive 4,000 years ago, when Egyptian pharaohs ruled.

More recent still was the extinction of one of the most astonishing of Pleistocene megafauna: the world's biggest bird, which also lived on an island humans overlooked. New Zealand's flightless moa, at 600 pounds, weighed twice as much an ostrich and stood nearly a yard taller. The first humans colonized New Zealand about two centuries before Columbus sailed to America. By the time he did, the last of 11 moa species was all but gone.

To Paul Martin, it's obvious. "Big animals were the easiest to track. Killing them gave humans the most food, and the most prestige." Within 100 miles of his Tumamoc Hill laboratory, past the Tucson jumble, are three of the 14 known Clovis kill sites. The richest of them, Murray Springs, strewn with Clovis spear points and dead mammoths, was found by two of Martin's students, Vance Haynes and Peter Mehringer. Its eroded strata, wrote Haynes, resembled "pages in a book that record the last 50,000 years of Earth history." Those pages contain obituaries of several extinct North American species: mammoth, horse, camel, lion, giant bison, and dire wolf. Adjacent sites add tapir, and two of the few megafauna that survive today: bear and bison.

Which raises a question: Why did they survive, if humans were slaughtering everything? Why does North America still have grizzlies, buffalo, elk, musk ox, moose, caribou, and puma, but not the other big mammals?

Polar bears, caribou, and musk ox inhabit regions where relatively few humans have ever lived—and those who did found fish and seals to be far easier prey. South of the tundra, where trees resume, lived bears and mountain lions, furtive and fleet creatures adept at hiding in forests or among boulders. Others, like *Homo sapiens*, entered North America around the time that the Pleistocene species departed. Today's plains buffalo are genetically closer to Poland's wisent than to the now-extinct giant bison that were killed at Murray Springs. After the giant bison were gone, the plains buffalo population exploded. Likewise, today's moose came from Eurasia after the American stag moose were extinguished.

Carnivores such as sabre-toothed tigers likely disappeared along with their prey. Some former Pleistocene residents—tapirs, peccaries, jaguars, and llamas—escaped farther south to forested refuges in Mexico, Central America, and beyond. Along with the die-off of the rest, this left huge niches to be filled, and eventually buffalo, elk, and company rushed in to fill them.

As Vance Haynes excavated Murray Springs, he found signs that drought had forced Pleistocene mammals to seek water—a cluster of footprints around one messy hole was clearly an attempt by mammoths to dig a well. There, they would have been easy pickings for hunters. In the layer just above the footprints is a band of black fossilized algae killed in a cold snap cited by many over-chill advocates—except, in paleontology's equivalent of a smoking gun, the mammoth bones all lie below it, not within it.

Yet one more clue that, had humans never existed, the descendants of these slaughtered mammoths would likely be around today: when their big prey vanished, so did Clovis people and their famous lithic points. With game gone and weather turned cold, perhaps they moved south. But within a matter of years, the Holocene warmed, and successors to the Clovis culture appeared, their smaller spear points tailored to smaller plains bison. An equilibrium of sorts was reached between these "Folsom people" and those remaining animals.

Had these succeeding generations of Americans absorbed a lesson from the gluttony of their ancestors who killed Pleistocene herbivores as if the supply were endless—until it crashed? Perhaps, although the existence of much of the Great Plains themselves is due to fires set by their descendants, the American Indians, both to concentrate game that browse, such as deer, in forest patches, and to create grassland for grazers like buffalo.

Later, as European diseases raced across the continent and nearly extirpated the Indians, the buffalo population surged and spread. They had almost reached Florida when white settlers heading west met them. After nearly all the buffalo were gone, save a few kept as curiosities, the white settlers took advantage of the plains that the Indians' ancestors had opened, and filled them with cattle.

From his hilltop laboratory, Paul Martin looks over a desert city that grew along a river, the Santa Cruz, which flowed north from Mexico. Camels, tapirs, native horses, and Columbian mammoths once foraged on its green floodplain. When descendants of the humans who eliminated them settled here, they built huts from mud and branches of riverbank cottonwood and willow—materials that quickly returned to the soil and the river when no longer needed.

With less game, the people learned to cultivate the plants they gathered, and they called the village that evolved Chuk Shon, a name that meant "flowing water." They mixed harvest chaff with river mud to form bricks, and this practice continued until mud adobes were supplanted by concrete after World War II. Not long after that, the advent of air-conditioning attracted so many people here that the river was sucked dry. They dug wells. When those dried, they dug deeper.

The Santa Cruz River's desiccated bed is now flanked by Tucson's civic center, which includes a convention hall whose jumbo concrete-and-steel-beam foundation seems like it should last at least as long as Rome's Coliseum. The tourists of some distant tomorrow might have a hard time finding it, however, because after today's thirsty humans are gone from Tucson and from the bloated Mexican border city of Nogales, Sonora, 60 miles south, eventually the Santa Cruz River will rise again. Weather will do what weather does, and from time to time Tucson and Nogales's dry river will be back in the business of building an alluvial plain. Silt will pour into the basement of the by-then-roofless Tucson Convention Center until it's buried.

What animals would live atop it is uncertain. Bison are long gone; in a world without people, the cows that replaced them won't last long without their attendant cowboys to discourage coyotes and mountain lions. The Sonoran pronghorn—a subspecies of that small, speedy Pleistocene relic, the last American antelope—verges on extinction in desert preserves not far from here. Whether there are enough left to replenish the breed before the coyotes finish them off is questionable, but possible.

Paul Martin descends Tumamoc Hill and drives his pickup truck west through a cactus-studded pass into the desert basin below. Before him lie mountains that are sanctuary to some of the last of North America's wildest creatures, including jaguar, bighorn sheep, and collared peccaries, locally known as *javelinas*. Many living specimens are on display just ahead at a famous tourist attraction, the Arizona–Sonora Desert Museum, which includes a zoo with subtle, naturally landscaped enclosures.

Martin's destination, a few miles shy of there, isn't subtle at all. The International Wildlife Museum was designed to replicate a French Foreign Legion fort in Africa. It houses the collection of a late millionaire big-game hunter, C. J. McElroy, who still holds many world records, including the world's biggest mountain sheep—a Mongolian argali—and the biggest jaguar, bagged in Sinaloa, Mexico. The special attractions here include a white rhino, one of 600 animals shot by Teddy Roosevelt during a 1909 African safari.

The museum's centerpiece is the faithfully reproduced 2,500-square-foot trophy room of McElroy's Tucson mansion, which bears the taxidermized spoils of a lifelong obsession with killing large mammals. Locally often derided as the "dead animal museum," for Martin on this night, it's perfect.

The occasion is the launch of his 2005 book, *Twilight of the Mammoths.* Just behind his audience rises a phalanx of grizzly and polar bears, frozen forever in mid-attack. Above the podium, its ears extended like gray spinnakers, is the trophy head of an adult African elephant. To either side, every breed of spiral horns found on five continents is represented. Pulling himself from his wheelchair, Martin slowly scans the hundreds of stuffed heads: bongo, nyala, bushbuck, sitatunga, greater and lesser kudu, eland, ibex, Barbary sheep, chamois, impala, gazelle, dik-dik, musk ox, cape buffalo, sable, roan, oryx, waterbuck, and gnu. Hundreds of pairs of glass eyes fail to return his moist blue gaze.

"I can't imagine a more appropriate setting," he says, "to describe what amounts to genocide. In my lifetime, millions of people slaughtered in death camps, from Europe's Holocaust to Darfur, are proof of what our species is capable of. My 50-year career has been absorbed by the extraordinary loss of huge animals whose heads don't appear on these walls. They were all exterminated, simply because it could be done. The person who put this collection together could have walked straight out of the Pleistocene."

He and his book conclude with a plea that his accounting of the Pleistocene mega-massacre be a cautionary lesson that stops us from perpetrating another that would be far more devastating. The matter is more complicated than a killer instinct that never relents until another species is gone. It involves acquisitive instincts that also can't tell when to stop, until something we never intended to harm is fatally deprived of something it needs. We don't actually have to shoot songbirds to remove them from the sky. Take away enough of their home or sustenance, and they fall dead on their own.

CHAPTER 6

The African Paradox

1. Sources

LUCKILY FOR THE world after humans, not all the big mammals are gone. A continent-sized museum, Africa, still holds a striking collection. Would they spread across the planet after we're gone? Could they replace what we finished off elsewhere, or even evolve to resemble those same lost creatures?

But first: If people come from Africa originally, why are elephants, giraffes, rhinos, and hippos even there at all? Why weren't they killed off, like 94 percent of Australia's large animal genera, most of them giant marsupials, or all the species that American paleontologists mourn?

Olorgesailie, site of the paleolithic tool factory discovered by Louis and Mary Leakey in 1944, is a dry yellow basin 45 miles southwest of Nairobi in the Eastern African Rift Valley. Much of it is dusted in white chalk from diatomaceous sediments, the stuff of swimming pool filters and kitty litter, composed of tiny fossilized exoskeletons of freshwater plankton.

The Leakeys saw that a lake had filled the Olorgesailie depression many times in prehistory, appearing in wet cycles and disappearing during drought. Animals came here to water, as did the toolmakers who pursued them. Ongoing digs now confirm that from 992,000 to 493,000 years ago, the lake's shore was inhabited by early humans. No actual hominid remains were found there until 2003, when archaeologists from the Smithsonian Institution and the National Museums of Kenya uncovered a single small skull, probably of *Homo erectus,* a predecessor of our own species.

What had been found, however, were thousands of stone hand-axes and cleavers. The most recent were designed for throwing: rounded on one end, with a point or double-faced edge on the other. Where protohumans at Olduvai Gorge, like *Australopithecus,* simply banged stones together until one chipped, these were flaked with techniques that could be duplicated, rock after rock. They are in every layer of human habitation here, meaning that people hunted and butchered game around Olorgesailie for at least half a million years.

Recorded history from civilization's Fertile Crescent beginnings to the present day has taken barely more than ⅟₁₀₀ th of the time that our ancestors lived in this one spot, grubbing plants and heaving sharpened stones at animals. There must have been a lot of prey to feed a growing predator population with awakening technological skill. Olorgesailie is cluttered with femurs and tibia, many smashed for their marrow. The quantities of stone tools surrounding the impressive remains of an elephant, a hippo, and an entire flock of baboons, suggest that the entire hominid community teamed up to kill, dismember, and devour their quarry.

Yet how is this possible if in less than a millennium human beings decimated America's supposedly richer Pleistocene megafauna? Surely Africa had even more people, and for a lot longer. If so, why does Africa still have its famous big-game menagerie? The flaked basalt, obsidian, and quartzite blades at Olorgesailie show that for a million years hominids could cut even an elephant's or a rhino's thick hide. Why aren't Africa's big mammals extinct as well?

Because here, humans and megafauna evolved together. Unlike the unsuspecting American, Australian, Polynesian, and Caribbean herbivores who had no inkling of how dangerous we were when we unexpectedly arrived, African animals had the chance to adjust as our presence increased. Animals growing up with predators learn to be wary of them, and they evolve ways to elude them. With so many hungry neighbors, African fauna have learned that massing in large flocks makes it harder for predators to isolate and catch a single animal, and assures that some are available to scout for danger while others feed. A zebra's stripes help it befuddle lions by getting lost in a crowded optical illusion. Zebras, wildebeest, and ostriches have forged a triple alliance on open savannas to combine the excellent ears of the first, the acute sense of smell of the second, and the sharp eyes of the third.

If these defenses worked every time, of course, the predators would go extinct. An equilibrium emerges: in a short sprint, the cheetah gets the gazelle; in a longer race, gazelles outlast the cheetah. The trick is to avoid becoming someone else's dinner long enough to breed replacements, or to breed often enough to insure that some replacements always survive. As a result, carnivores like lions often end up harvesting the sickest, oldest, and weakest. That was what early humans did as well—or, like hyenas, at first we probably did something even easier: we ate the carrion left by some more adept hunter.

Equilibrium dissolves, however, when something changes. The genus *Homo*'s burgeoning brain spawned inventions that challenged herbivore defense strategies: tight flocks, for example, increased the odds that a thrown hand-axe would actually connect with a target. Many species found in Olorgesailie sediments, in fact, are now extinct, including a horned giraffe, a giant baboon, an elephant with down-curved tusks, and a hippopotamus even beefier than today's. It isn't clear, however, that humans drove them to extinction.

This, after all, was the mid-Pleistocene—a time when 17 ice ages and their interregna yanked global temperatures up and down and alternately soaked or parched any land that wasn't frozen solid. The Earth's crust squeezed and relaxed under the shifting weight of ice. The Eastern African Rift widened and volcanoes blew, including one that periodically bombed Olorgesailie with ashes. After two decades of studying Olorgesailie's strata, Smithsonian archaeologist Rick Potts began to notice that certain persistent species of plants and animals typically survived periods of climatic and geologic upheaval.

One of these was us. At Lake Turkana, a Rift lake shared by Kenya and Ethiopia, Potts tallied a rich trove of our ancestors' remains and realized that whenever climate and environmental conditions grew unruly, early species of *Homo* outnumbered, and finally displaced, even earlier hominids. Adaptability is the key to who is fittest, one species' extinction being another's evolution. In Africa, megafauna fortunately evolved their own adaptable forms right along with us.

That is fortunate for us, too, because to picture how the world was before us—as our basis for understanding how the world may evolve after us—Africa is our most complete bank of living genetic heritage, filled with entire families and orders of animals that were sacked elsewhere. Some actually are *from* elsewhere: when North Americans stand in the open sunroof of a safari jeep in the Serengeti, stunned by the vastness of a herd of zebras, they're seeing descendants of American species that herd over Asian and Greenland-European land bridges, but are now lost to their own continent. (That is, until Columbus reintroduced *Equus* after a hiatus of 12,500 years; before that, some horse species that flourished in America were probably also striped.)

If Africa's animals evolved learning to avoid human predators, how would the balance swing with humans gone? Are any of its megafauna so adapted to us that some subtle dependence or even symbiosis would be lost along with the human race, in a world without us?

The high, cold Aberdares moors in central Kenya have discouraged human settlers, though people must have always made pilgrimages to this source. Four rivers are born here, heading in four directions to water Africa below, plunging along the way from basalt overhangs into deep ravines. One of these waterfalls, the Gura, arcs through nearly 1,000 feet of mountain air before being swallowed by mist and tree-sized ferns.

In a land of megafauna, this is an alpine moor of mega*flora.* Except for a few pockets of rosewood, it is above the tree line, occupying a long saddle between two 13,000-foot peaks that form part of the Rift Valley's eastern wall, just below the equator. Treeless—yet giant heather rises 60 feet here, dripping curtains of lichen. Groundcover lobelia turns into columns eight feet high, and even groundsel, usually just a weed, mutates into 30-foot trunks with cabbage tops, growing amid massive grass tussocks.

Small wonder that the descendants of early *Homo* who climbed out of the Rift and eventually became Kenya's highland Kikuyu tribe figured that this was where Ngai—God—lived. Beyond the wind in the sedges and the tweep of wagtails, it's sacredly quiet. Rills lined with yellow asters flow soundlessly across spongy, hummocked meadows, so rain-logged that streams appear to float. Eland—Africa's biggest antelopes, seven feet tall and 1,500 pounds, their helix horns a yard long, their numbers dwindling—seek refuge at these freezing heights. The moor is too high for most game, though, except for waterbucks and hidden lions who await them in fern forests along the plunge pools.

At times elephants appear, babies following a big tusker as she stomps through purple clover and smashes giant thickets of St. John's wort in pursuit of her daily 400 pounds of forage. Fifty miles east of Aberdares, across a flat valley, elephants have been spotted near the snow line of Mount Kenya's 17,000-foot spire. Far more adaptable than their late woolly mammoth cousins, individual African elephants once could be tracked by trails of dung leading from Mount Kenya or the cold Aberdares down to Kenya's Samburo desert, an elevation drop of two miles. Today, the din of humanity interrupts the corridors linking those three habitats. The elephant populations of Aberdares, Mount Kenya, and Samburo have not seen each other for decades.

Below the moor, a 1,000-foot band of bamboo circles the Aberdares Mountains, sanctuary to the nearly extinct bongo, another of Africa's striped camouflagees. In bamboo so dense it discourages hyenas and even pythons, the spiral-horned bongo's only predator is unique to the Aberdares: the seldom-seen melanistic, or black, leopard. The brooding Aberdares rain forest is also home to a black serval and a black race of the African golden cat.

It's one of the wildest places left in Kenya, with camphor, cedar, and croton trees so thick with lianas and orchids that 12,000-pound elephants easily hide here. So does the most imperiled of all African species: the black rhino. About 400 remain in Kenya, down from 20,000 in 1970, the rest poached for horns that bring $25,000 each in the Orient for alleged medicinal properties, and in Yemen for use as ceremonial dagger handles. The estimated 70 Aberdares black rhinos are the only ones in their original wild habitat.

Humans once hid here, too. During colonial times, the well-watered, volcanic Aberdares slopes belonged to British tea and coffee growers who alternated their plantations with sheep and cattle ranches. The agricultural Kikuyu were reduced to sharecropping plots called *shambas* on their now-conquered land. In 1953, under the cover of the Aberdares forest, they organized. Surviving on wild figs and the brown speckled trout stocked by the British in Aberdares streams, Kikuyu guerrillas terrorized white landowners in what became known as the Mau Mau Rebellion. The Crown brought divisions from England and bombed the Aberdares and Mount Kenya. Thousands of Kenyans were killed or hung. Barely 100 British died, but by 1963 a negotiated truce had inexorably led to majority rule, which became known in Kenya as *uhuru*—independence.

Today, the Aberdares is an example of that wobbly kind of pact that we humans have struck with the rest of nature known as a national park. It is haven to rare giant forest hogs and the smallest antelopes—jackrabbit-sized suni—and to golden-winged sunbirds, silvery-cheeked hornbills, and incredible scarlet-and-beyond-blue Hartlaub's turacos. The black-and-white colobus monkey, whose bearded visage surely shares genes with Buddhist monks, dwells in this primal forest, which sweeps in all directions down the slopes of the Aberdares . . .

. . . until it stops at an electric fence. Two hundred kilometers of galvanized wire, pulsing 6,000 volts, now encircle Kenya's greatest water catchment. Electrified mesh rises seven feet above ground and is buried three feet beneath it, its posts hot-wired to keep baboons, vervet monkeys, and ringed-tailed civets off them. Where it crosses a road, electrified arches allow vehicles to pass, but dangling live wires deter vehicle-sized elephants from doing the same.

It is a fence to protect animals and people from each other. On either side lies some of the best soil in Africa, planted in forest above and in corn, beans, leeks, cabbage, tobacco, and tea below. For years, incursions went in both directions. Elephants, rhinos, and monkeys invaded and uprooted fields by night. Burgeoning Kikuyu populations snuck farther up the mountain, felling 300-year-old cedars and *podo* conifers as they advanced. By 2000, nearly one-third of the Aberdares was cleared. Something had to be done to keep trees locked in place, to keep enough water transpiring through leaves and raining back into Aberdares rivers, to keep them flowing to thirsty cities like Nairobi, and to keep hydroelectric turbines spinning and Rift lakes from disappearing.

Hence, the world's longest electric barricade. By then, however, the Aberdares had other water problems. In the 1990s, a deep new drain had opened at its skirts, cloaked innocently in roses and carnations, as Kenya passed Israel to become Europe's biggest provider of cut flowers, which now exceed coffee as its main source of export income. This fragrant turn of fortune, however, incurs a debt that may keep compounding long after flower lovers are no longer around.

A flower, like a human, is two-thirds water. The amount of water a typical floral exporter therefore ships to Europe each year equals the annual needs of a town of 20,000 people. During droughts, flower factories with production quotas stick siphons into Lake Naivasha, a papyrus-lined, freshwater bird and hippo sanctuary just downstream from the Aberdares. Along with water, they suck up entire generations of fish eggs. What trickles back whiffs of the chemical trade-off that keeps the bloom on a rose flawless all the way to Paris.

Lake Naivasha, however, doesn't look quite so alluring. Phosphates and nitrates leached from flower greenhouses have spread mats of oxygen-choking water hyacinth across its surface. As the lake level drops, water hyacinth—a South American perennial that invaded Africa as a potted plant—crawls ashore, beating back the papyrus. The rotting tissues of hippo carcasses reveal the secret to perfect bouquets: DDT and, 40 times more toxic, Dieldrin—pesticides banned in countries whose markets have made Kenya the world's number-one rose exporter. Long after humans and even animals or roses go, Dieldrin, an ingeniously stable, manufactured molecule, may still be around.

No fence, not even one packing 6,000 volts, can ultimately contain the animals of the Aberdares. Their populations will either burst the barriers or wither as their gene pools shrink, until a single virus snuffs an entire species. If humans are snuffed first, however, the fence will stop dispensing jolts. Baboons and elephants will make an afternoon fete of the grains and vegetables in the surrounding Kiyuku *shambas*. Only coffee stands a chance to survive; wildlife don't crave caffeine very much, and the arabica strains brought long ago from Ethiopia liked central Kenya's volcanic soils so much they've gone native.

Wind will shred the polyethylene greenhouse covers, their polymers embrittled by equatorial ultraviolet rays whose potency is abetted by the flower industry's favorite fumigant, methyl bromide, the most potent ozone destroyer of all. The roses and carnations, addicted to chemicals, will starve, although water hyacinth may outlast everything. The Aberdares forest will pour through the deactivated fence, repossessing *shambas* and overrunning an old colonial relic below, the Aberdares Country Club, its fairways currently kept trimmed by resident warthogs. Only one thing stands in the forest's way from reconnecting the wildlife corridors up to Mount Kenya and down to the Samburo desert: a ghost of the British Empire, in the form of eucalyptus groves.

Among the myriad species loosed on the world by humans that have surged beyond control, eucalyptus joins ailanthus and kudzu as encroachers that will bedevil the land long after we've departed. To power steam locomotives, the British often replaced slow-maturing tropical hardwood forests with fast-growing eucalyptus from their Australian Crown colonies. The aromatic eucalyptus oils that we use to make cough medicine and to disinfect household surfaces kill germs because in larger doses they're toxins, meant to chase off competitive plants. Few insects will live around eucalyptus, and with little to eat, few birds nest among them.

Lusty drinkers, eucalypti go wherever there's water, such as along *shamba* irrigation ditches, where they've formed tall hedgerows. Without people, they'll aim to colonize deserted fields, and they'll have a head start on the native seeds blowing down the mountain. In the end, it may take a great natural African lumberjack, the elephant, to blaze a trail back to Mount Kenya and expel the last British spirits from the land for good.

2. Africa After Us

In an Africa without humans, as elephants push above the equator through Samburo and then beyond the Sahel, they may find a Sahara Desert in northward retreat, as desertification's advance troops—goats—become lunch for lions. Or, they may collide with it, as temperatures rising on a wave of a human legacy, elevated atmospheric carbon, quicken its march. That the Sahara has lately advanced so rapidly and alarmingly—in places, two to three miles per year—owes to unfortunate timing.

Only 6,000 years ago, what is now the world's largest nonpolar desert was green savanna. Crocodiles and hippos wallowed in plentiful Sahara streams. Then Earth's orbit underwent one of its periodic readjustments. Our tilted axis straightened not even half a degree, but enough to nudge rain clouds around. That alone was not sufficient to turn grasslands to sand dunes. But the coincidence of human progress tipped what was becoming an arid shrubland over a climatic edge. During two previous millennia, in North Africa, *Homo sapiens* had gone from hunting with spears to growing Middle Eastern grains and raising livestock. They mounted their belongings, and themselves, on newly tamed descendants of an American ungulate that luckily emigrated before its cousins back home perished in a megafaunal holocaust: the camel.

Camels eat grass; grass needs water. So did their masters' crops, whose bounty begat a population boom of humans. More humans needed more herds, pasture, fields, and more water—all at just the wrong time. No one could have known that the rains had shifted. So people and their flocks ranged farther and grazed harder, assuming that the weather would return to what it had been, and that everything would grow back the way it was.

It didn't. The more they consumed, the less moisture transpired skyward and the less it rained. The result was the hot Sahara we see today. Only it used to be smaller: Over this past century, the numbers of Africa's humans and their animals have been rising, and now temperatures are, too. This leaves the precarious sub-Saharan tier of Sahel countries at the brink of sliding into the sand.

Farther south, equatorial Africans have herded animals for several thousand years and hunted them even longer, yet between wildlife and humans there was actually mutual benefit: As pastoralists such as Kenya's Maasai shepherded cattle among pastures and water holes, their spears ready to discourage lions, wildebeest tagged along to take advantage of the predator protection. They, in turn, were followed by their zebra companions. The nomads economized by eating meat sparingly, learning to live on their flocks' milk and blood, which they drew by carefully tapping and staunching their cattle's jugular veins. Only when drought reduced fodder for their herds did they fall back on hunting, or trade with bushmen tribes that still lived off game.

This balance among humans, flora, and fauna first began to shift when humans became prey themselves—or rather, commodities. Like our kin the chimpanzees, we'd always murdered one another over territory and mates. But with the rise of slavery, we were reduced to something new: an export crop.

The mark that slavery left on Africa can be seen today in southeastern Kenya, in brushy country known as Tsavo, an eerie landscape of lava-flows, flat-topped tortilis acacias, myrrh, and baobab trees. Because Tsavo's tsetse flies discouraged cattle herding, it remained a hunting ground for Waata bushmen. Their game included elephant, giraffe, cape buffalo, assorted gazelles, klipspringer, and another striped antelope: the kudu, its horns corkscrewing for an amazing six feet.

The destination for black slaves in East Africa was not America, but Arabia. Until the mid-19th century, Mombasa, on Kenya's coast, was the shipping port for human flesh, the end of a long line for Arab slavers who captured their merchandise at gunpoint in central African villages. Caravans of slaves marched barefoot down from the Rift, herded by armed captors mounted on donkeys. As they descended into Tsavo, the heat rose and tsetse flies swarmed. Slavers, shooters, and whichever prisoners had survived the journey made for a fig-shaded oasis, Mzima Springs. Its artesian pools, filled with terrapins and hippopotamuses, were refreshed daily by 50 million gallons of water upwelling from porous volcanic hills 30 miles away. For days slave caravans paused here, paying Waata bow-hunters to replenish their stores. The slave route was also the ivory route, and every elephant encountered was harvested. As demand for ivory grew, its price outstripped that of slaves, who became chiefly valued as ivory porters.

Near Mzima Springs, the water outcropped again, forming the Tsavo River, which eventually led to the sea. With shady groves of fever trees and palms, this route was irresistible, but the price was often malaria. Jackals and hyenas followed the caravans, and Tsavo's lions developed reputations as maneaters by dining on dying slaves left behind.

Until the late 19th century, when the British ended slavery, thousands of elephants and humans perished along the ivory-slave route between the central plains and Mombasa's auction block. As the slave trail closed, construction commenced on a railroad between Mombasa and Lake Victoria, a source of the Nile, critical to British colonial control. Tsavo's hungry lions gained international fame for devouring railway workers, sometimes leaping aboard trains to corner them. Their appetites became the stuff of legend and movies, which usually failed to mention that their hunger owed to a scarcity of other game, slaughtered to feed a 1,000-year cavalcade of enslaved human cargo.

After slavery and rail construction, Tsavo was an abandoned, empty country. Without people, its wildlife began creeping back. Briefly, so did armed humans. From 1914 to 1918, Britain and Germany, which had previously agreed to carve up much of Africa between them, were fighting a Great War for reasons that seemed even murkier in Africa than in Europe. A battalion of German colonists from Tanganyika—today, Tanzania—blew up the British Mombasa-Victoria railroad on several occasions. The two sides engaged each other amid palms and fever trees along the Tsavo River, living on bush meat and dying of malaria as much as from bullets, but bullets having the usual disastrous repercussions for wildlife.

Again, Tsavo was emptied. Again, in the absence of humans, it filled with animals. Sandpaper trees laden with yellow saucerberries overgrew the World War I battlefields, hosting families of baboons. In 1948, stating that people had no other use for it, the Crown declared Tsavo, one of human history's busiest trade routes, a wilderness refuge. Two decades later, its elephant population was 45,000—one of Africa's biggest. That, however, was not to last.

As the white single-engine Cessna takes off, one of the Earth's most incongruous sights unfolds beneath its wings. The wide savanna below is Nairobi National Park, where elands, Thomson's gazelles, cape buffalo, hartebeest, ostriches, white-bellied bustards, giraffes, and lions live jammed against a wall of blocky high-rises. Behind that gray urban facade begins one of the world's largest, poorest slums. Nairobi is only as old as the railroad that needed a depot between Mombasa and Victoria. One of the youngest cities on Earth, it will likely be among the first to go, because even new construction here quickly begins to crumble.

On its opposite end, Nairobi National Park is unfenced. The Cessna passes its unmarked boundary, crossing into a gray plain dotted with morning-glory trees. Through here, the park's migrating wildebeest, zebra, and rhinos follow seasonal rains along a corridor lately pinched by maize fields, flower farms, eucalyptus plantations, and sprawling new fenced estates with private wells and conspicuous large houses. Together, these may turn Kenya's oldest national park into yet another wildlife island. The corridor isn't protected; with real estate outside of roiling Nairobi becoming increasingly attractive, the best option, in the opinion of the Cessna's pilot, David Western, is for the government to pay owners to let animals cross their property. He's helped with negotiations, but he's not hopeful. Everyone fears elephants squashing their gardens, or worse.

Counting elephants is David Western's project today—something he has done continually for nearly three decades. Raised in Tanzania, son of a British big-game hunter, as a boy he often hiked alongside his gun-toting father for days without seeing another human. The first animal he shot was his last; the look in the dying warthog's eyes cooled any further passion to hunt. After an elephant fatally gored his father, his mother took her children to the comparative safety of London. David stayed through university studies in zoology, then returned to Africa.

An hour southeast of Nairobi, Kilimanjaro appears, its shrinking snowcap dripping butterscotch under the rising sun. Just before it, verdant swamps burst from a brown alkaline basin, fed by springs from the volcano's rainy slopes. This is Amboseli, one of Africa's smallest, richest parks, an obligatory pilgrimage for tourists hoping to photograph elephants silhouetted against Kilimanjaro. That used to be a dry-season event, when wildlife would pack into Amboseli's marshland oasis to survive on cattails and sedges. Now they're always here. "Elephants aren't supposed to be sedentary," Western mutters as he passes over dozens of females and calves wading not far from a pod of mucking hippos.

From high above, the plain surrounding the park seems infected by giant spores. These are *bomas*: rings of mud-and-dung huts belonging to Maasai pastoralists, some occupied, some abandoned and melting back into the earth. A defensive ring of stacked, thorny acacia branches encircles each. The bright green patch in every compound's center is where the nomadic Maasai keep cattle safe from predators at night before moving their herds and families to the next pasture.

As Maasai move out, elephants move in. Since people first brought cattle down from northern Africa after the Sahara dried, a choreography has evolved featuring elephants and livestock. After cattle chew savanna grasses down, woody shrubs invade. Soon they're tall enough for elephants to munch, using their tusks to strip and eat bark, knocking trees over to reach their tender canopies, clearing the way for grass to return.

As a graduate student, David Western sat atop an Amboseli hill, counting cows led to graze by Maasai herders as elephants plodded in the opposite direction to browse. The census he began here of cattle, elephants, and people has never stopped during his subsequent careers as Amboseli park director, head of the Kenya Wildlife Service, and founder of the nonprofit African Conservation Centre, which works to preserve wildlife habitats by accommodating, not banning, humans who have traditionally shared them.

Dropping to 300 feet, he begins flying wide, clockwise circles, banked at a 30° angle. He tallies a ring of dung-plastered huts—one hut per wife: some wealthy Maasai have as many as 10 wives. He calculates the approximate number of inhabitants, and notes 77 cattle on his vegetation map. What looked from above like blood drops on a green plain turns out to be the Maasai herders themselves: tall, lithe, dark men in traditional red plaid shoulder cloaks—traditional, at least, since the 19th century, when Scottish missionaries distributed tartan blankets that Maasai herdsmen found both warm and light enough to carry as they followed their herds for weeks.

"The pastoralists," Western shouts over the engine noise, "have become a surrogate migratory species. They behave much like wildebeest." Like the wildebeest, Maasai herd their cows into short-grass savannas during wet seasons and bring them back to water holes when the rains stop. Over a year, Amboseli's Maasai live in an average of eight settlements. Such human movement, Western is convinced, has literally landscaped Kenya and Tanzania to the benefit of wildlife.

"They graze their cattle and leave behind woodland for elephants. In time, elephants create grassland again. You get a patchy mosaic of grass, woods, and shrublands. That's the whole reason for the savanna's diversity. If you only had woodlands or grasslands, you would only support woodland species or grassland species."

In 1999, Western described this to paleoecologist Paul Martin, father of the Pleistocene overkill extinction theory, while driving through southern Arizona en route to see where Clovis people finished off local mammoths 13,000 years earlier. Since that time, the American Southwest had evolved without big herbivore browsers. Martin gestured at the tangle of mesquite sprouting on public lands that ranchers leased, which they were always begging permission to burn. "Do you think this could work as elephant habitat?" he asked.

At the time, David Western laughed. But Martin persisted: How would African elephants do in this desert? Would they be able to ascend the craggy granite mountain ranges to find water? Might Asian elephants do better, since they were more closely related to mammoths?

"It's surely better than using a bulldozer and herbicides to get rid of mesquite," Western agreed. "Elephants would do it a lot more cheaply and simply, and they also spread manure around for grass seedlings."

"Exactly," said Martin, "what mammoths and mastodons did."

"Sure," Western replied. "Why not use an ecological surrogate species if you don't have the original one there?" Ever since, Paul Martin has been campaigning to return elephants to North America.

Unlike Maasai, however, American ranchers aren't nomads who regularly vacate niches for elephants to use. Increasingly, though, Maasai and their cows are also staying put. The barren, overgrazed ground ringing Amboseli National Park testifies to the result. When light-haired, fair-skinned David Western, of medium height, chats in Swahili with 7-foot, ebony Maasai herdsmen, the contrast dissolves in long-standing mutual regard. Land subdivision has long been their common foe. But with developers and immigrants from rival tribes putting up fences and staking claims, the Maasai have no choice but to seek title and cling to their lands as well. The new human-use pattern reshaping Africa may not be easily obliterated when humans are gone, says Western.

"It's a bipolar situation. When you force elephants inside a park, and you graze cattle outside, you get two very different habitats. Inside, you lose all your trees, and it becomes grasslands. Outside, it becomes thick bushlands."

During the 1970s and 1980s, elephants learned the hard way to stay where they're safe. Unwittingly, they lumbered into a global collision between deepening African poverty, which in Kenya was yoked to the planet's highest birthrate, and the boom that spawned the so-called Asian economic tigers, which unleashed a craving in the Far East for luxuries. These included ivory; the desire for it outstripped even the lust that once financed centuries of slavery.

As the price, $20 per kilo, rose by a factor of 10, ivory poachers turned places like Tsavo into a trash heap of tuskless carcasses. By the 1980s, more than half of Africa's 1.3 million elephants were dead. Only 19,000 were left in Kenya, packed into sanctuaries such as Amboseli. International ivory bans and shoot-to-kill orders for poachers calmed but never eradicated the carnage, especially the slaughter of elephants outside parks on the pretext of defending crops or people.

The fever tree acacias that once lined Amboseli's swamps are now gone, downed by overcrowded pachyderms. As parks become treeless plains, desert creatures like gazelles and oryx replace browsers like giraffes, kudus, and bushbuck. It is a man-made replica of extreme drought, such as Africa knew during ice ages, when habitats shriveled and creatures crammed into oases. Africa's megafauna made it through those bottle-necks, but David Western fears what may happen to them in this one, stranded on island refuges in a sea of settlements, subdivisions, exhausted pastures, and factory farms. For thousands of years, migratory humans were their escorts across Africa: nomads and their herds taking what they needed and moving on, leaving nature even richer in their wake. But now such human migration is coming to a close. *Homo sedentarian* has flipped that scenario. Food now migrates to us, along with luxury goods and other consumables that never existed through most of human history.

Unlike anywhere else on Earth—save Antarctica, where people never settled—Africa alone has not suffered a major wildlife extinction. "But intensified agriculture and high human population," Western worries, "mean that we're looking at one now." The balance that evolved between humans and wildlife in Africa has tipped out of control: too many people, too many cows, too many elephants stuffed into too few spaces by too many poachers. The hope that sustains David Western lies in knowing that some of Africa is still as it was, before we evolved into a keystone species potent enough to push even elephants around.

If there were no people left, he believes, Africa, which has been occu-pied by humans longer than any other place would paradoxically revert to the purest primeval state on Earth. With so much wildlife grazing and browsing, Africa is the only continent where exotic plants haven't escaped suburban gardens to usurp the countryside. But Africa after people would include some key changes.

Once, North African cattle were wild. "But after thousands of years with humans," says Western, "they've been selected for a gut like an over-sized fermentation vat to eat huge amounts of forage during the day, be-cause they can't graze at night. So now they're not very quick. Left on their own, they'd be rather vulnerable prime beef."

And a lot of it. Cattle now account for more than half the live weight of African savanna ecosystems. Without Maasai spears to protect them, they would provide an orgy for binging lions and hyenas. Once cows were gone, there would be more than double the feed for everything else. Shading his eyes, Western leans against his Jeep and calculates what the new numbers would mean. "A million and a half wildebeest can take out grass just as effectively as cattle. You'd see much tighter interaction between them and elephants. They would play the role the Maasai refer to when they say that 'cattle grow trees, elephants grow grass.'"

As for elephants without people: "Darwin estimated 10 million elephants in Africa. That was actually quite close to what was here before the big ivory trade." He turns to look at the female herd sloshing in the Amboseli swamp. "At the moment we have half a million."

No people and 20 times more elephants would restore them as the undisputed keystone species in a patchwork mosaic African landscape. By contrast, in North and South America, for 13,000 years nearly no creatures except insects have eaten tree bark and bushes. After mammoths died, enormous forests would spread unless farmers cleared them, ranchers burned them, peasants cut them for fuel, or developers bulldozed them. Without humans, American forests represent vast niches awaiting any herbivore big enough to extract their woody nutrients.

3. Insidious Epitaph

Partois ole Santian heard the story often when he was growing up, wandering with his father's cows west of Amboseli. He listens respectfully as Kasi Koonyi, the gray old man living with his three wives in a *boma* in Maasai Mara, where Santian now works, tells it again.

"In the beginning, when there was only forest, Ngai gave us bushmen to hunt for us. But then the animals moved away, too far to be hunted. The Maasai prayed to Ngai to give us an animal that wouldn't move away, and He said wait seven days."

Koonyi takes a hide strap and holds one end of it skyward, to demonstrate a ramp sweeping down to Earth. "Cattle came down from heaven, and everyone said, 'Look at that! Our god is so kind, he sent us such a beautiful beast. It has milk, beautiful horns, and different colors. Not like wildebeest or buffalo, with only one color.'"

At this point, the story gets sticky. The Maasai claim all the cattle are meant for them, and kick the bushmen out of their *bomas*. When the bushmen ask Ngai for their own cattle to feed themselves, He refuses, but offers them the bow and arrow. "That's why they still hunt in the forests instead of herding like we Maasai."

Koonyi grins, his wide eyes glowing red in afternoon sun that flashes off the pendulous, cone-shaped bronze earrings that stretch his lobes chin-ward. The Maasai, he explains, figured out how to burn trees to create sa-vannas for their herds; the fires also smoked out malarial mosquitoes. Santian gets his drift: When humans were mere hunter-gatherers, we weren't much different from any other animal. Then we were chosen by God to became pastoralists, with divine dominion over the best animals, and our blessings grew.

The trouble is, Santian also knows, the Maasai didn't stop there.

Even after white colonials took so much grazing land, nomadic life had still worked. But Maasai men each took at least three wives, and as each wife bore five or six children, she needed about 100 cows to support them. Such numbers were bound to catch up with them. In Santian's young life-time, he has seen round *bomas* become keyhole-shaped as Maasai appended fields of wheat and corn and began to stay in one place to tend them. Once they became agriculturalists, everything began to change.

Partois ole Santian, who grew up in a modernizing Maasai generation with the option of studying, excelled in sciences, learned English and French, and became a naturalist. At 26, he became one of a handful of Africans to earn silver certification from the Kenya Professional Safari Guide Association—the highest level. He found work with an ecotourism lodge in the Kenyan extension of Tanzania's Serengeti Plain, Maasai Mara, a park combining an animals-only reserve with mixed conservation areas where Maasai, their herds, and wildlife might coexist as they always have. The red oat-grass Maasai Mara plain, dotted with desert date and flat-topped acacias, is still as splendid a savanna as any in Africa. Except that the most predominant animal grazing here is now the cow.

Often, Santian ties leather shoes on his long legs and climbs Kileleoni Hill, the highest point in the Mara. It is still wild enough to find impala carcasses hanging from tree limbs where leopards have stored them. From the top, Santian can look 60 miles south into Tanzania and the immense green-grass sea of the Serengeti. There, honking wildebeest mill in huge

June flocks that will soon merge like floodwaters and burst across the border, bounding through rivers that boil with crocodiles awaiting their annual northward migration, with lions and leopards dozing above in the tortilis trees, needing only to roll over to make a kill.

The Serengeti has long been an object of Maasai bitterness: half a million square kilometers from which they were swept away in 1951, for a theme park cleansed of a keystone species, *Homo sapiens,* to humor Hollywood-bred tourist delusions of Africa as wilderness primeval. But Maasai naturalists like Santian are now grateful for it: the Serengeti, blessed with perfect volcanic soils for grassland, is gene bank to the richest concentration of mammals on Earth, a source from which species might one day radiate and repopulate the rest of the planet, if it comes to that. Huge as it is, however, naturalists worry about how the Serengeti will maintain all those uncountable gazelles, let alone elephants, if everything surrounding it turns into farms and fences.

There isn't enough rain to change all the savanna into arable farmland. But that hasn't stopped the Maasai from multiplying. So far married to only one woman, Partois ole Santian decided to stop right there. But Noonkokwa, the childhood girlfriend he wed upon completing his traditional warrior training, was horrified to learn that she might be in this marriage alone, with no female companions.

"I'm a naturalist," he explained to her. "If all the wildlife habitat disappeared, I'd have to farm." Before subdividing began, Maasai considered farming beneath the dignity of men chosen by God to pastor cattle. They wouldn't even break sod to bury someone.

Noonkokwa understood. But she was still a Maasai woman. They compromised at two wives. But she still wanted six children. He was hoping to hold it to four; the second wife, of course, would want some, too.

Only one thing, too terrible to contemplate, might slow all this proliferating before all the animals go extinct. The old man, Koonyi, had said it himself. "The end of the Earth," he called it. "In time, AIDS will wipe out humans. The animals will take it all back."

AIDS isn't yet the nightmare for the Maasai that it has become for sedentary tribes, but Santian saw how it could be soon. Once, Maasai only traveled on foot through savannas with their cows, spear in hand. Now some go to towns, sleep with whores, and spread AIDS on their return. Even worse are the lorry drivers who now show up twice a week, bringing gasoline for the pickups, motor scooters, and tractors that Maasai farmers purchase. Even young uncircumcised girls are getting infected.

In non-Maasai areas, such as up at Lake Victoria, where the Serengeti animals migrate each year, coffee growers too sick with AIDS to groom their plants have turned to growing easy staples like bananas, or cutting trees to make charcoal. Coffee bushes, now feral, are 15 feet high, beyond rehabilitation. Santian has heard people say they don't care anymore, there's no cure, so they won't stop having children. So orphans now live with a virus instead of with parents, in villages where the adults have been all but wiped out.

Houses with no one left alive are collapsing. Mud-stick huts with dung roofs have melted away, leaving only half-finished houses of brick and cement begun by traders with money made from driving their lorries. Then they got sick, and gave their money to herbalists to cure them and their girlfriends. Nobody got well, and construction never resumed. The herbalists got all the money, then got sick themselves. In the end, the traders died, the girlfriends died, the medicine men died, and the money vanished; all that remains are roofless houses with acacias growing in the middle, and infected children who sell themselves to survive until they die early.

"It's wiping out a generation of future leaders," Santian had replied to Koonyi that afternoon, but the old Maasai figured that future leaders wouldn't matter much with animals back in charge.

The sun rolls along the Serengeti Plain, filling the sky with iridescence. As it falls over the edge, blue twilight settles on the savanna. The day's remaining warmth floats up the side of Kileleoni Hill and dissolves into the dusk. The chilly updraft that follows carries the screech of baboons. Santian pulls his red-and-yellow tartan *shuka* tighter.

Could AIDS be the animals' final revenge? If so, *Pan troglodytes,* our chimpanzee siblings in central Africa's womb, are accessories to our undoing. The human immunodeficiency virus that infects most people is closely related to a simian strain that chimps carry without getting sick. (The less-common HIV-II is similar to a form carried by rare mangabey monkeys found in Tanzania.) Infection probably spread to humans through bush meat. On encountering the 4 percent of our genes that differ from the genes of our closest primate relations, the virus mutated lethally.

Had moving to the savanna somehow made us biochemically more vulnerable? Santian can identify every mammal, bird, reptile, tree, and spider, and most flowers, visible insects, and medicinal plants in this ecosystem, but some subtle genetic differences escape him—and everyone searching for an AIDS vaccine as well. The answer may be in our brain, since brain size is where humans differ significantly from chimps and bonobos.

Another burst of yakking from the baboon troop drifts up from below. Probably they're harassing the leopard who hung the impala meat. Interesting how male baboons vying for alpha status have learned to maintain a truce long enough to cooperate in discouraging leopards. Baboons also have the largest brain of any primate after *Homo sapiens,* and are the only other primate that adapted to living in savannas as forest habitats shrank.

If the dominant ungulates of the savanna—cattle—disappear, wildebeest will expand to take their place. If humans vanish, will baboons move into ours? Has their cranial capacity lay suppressed during the Holocene because we got the jump on them, being first out of the trees? With us no longer in their way, will their mental potential surge to the occasion and push them into a sudden, punctuated evolutionary scramble into every cranny of our vacant niche?

Santian rises and stretches. A new moon rocks toward the equatorial horizon, its points curving upward like a bowl for silvery Venus to settle inside. The Southern Cross and Milky Way assume their places. The air smells like violets. Up here, Santian hears wood owls, like those he knew in his boyhood until the forests around their *bomas* turned to wheat fields. If human crops revert to a mosaic of woods and grassland, and if baboons fill our keystone slot, would they be satisfied to dwell in pure natural beauty?

Or would curiosity and sheer narcissistic delight in their unfolding powers eventually push them and their planet to the brink, too?

PART II

❦

CHAPTER 7

What Falls Apart

IN THE SUMMER of 1976, Allan Cavinder got a call he wasn't expecting. The Constantia Hotel in Varosha was reopening under a new name after standing vacant for nearly two years. A lot of electrical work was needed—was he available?

This was a surprise. Varosha, a resort on the eastern shore of the Mediterranean island of Cyprus, had been off-limits to everyone since war fractured the country two years earlier. The actual fighting had lasted only a month before the United Nations stepped in to broker a messy truce between Turkish and Greek Cypriots. A no-man's zone called the Green Line was drawn wherever opposing troops found themselves at the exact moment of the cease-fire. In the capital, Nicosia, the Green Line wandered like a drunk among bullet-scarred avenues and houses. On narrow streets where hand-to-hand combat had been underway between enemies jabbing bayonets from facing balconies, it was little more than 10 feet wide. In the country, it broadened to five miles. Turks now lived to the north and Greeks to the south of a weedy UN-patrolled strip, refuge to hares and partridges.

When the war broke out in 1974, much of Varosha was barely two years old. Strung along a sand crescent south of the deep-water port of Famagusta, a walled city dating to 2000 BC, Varosha had been developed by Greek Cypriots as Cyprus's Riviera. By 1972, tall hotels extended three uninterrupted miles along Varosha's golden beach, backed by blocks of shops, restaurants, cinemas, vacation bungalows, and employee housing.

The location had been chosen for the gentle, warm waters on the island's wind-sheltered eastern coast. The sole flaw was the decision, repeated by nearly every beachfront high-rise, to build as close to shore as possible. Too late, they realized that once the sun peaked at noon, the beach would lie in a shadow cast by the palisade of hotels.

There wasn't much time to worry about that, though. In the summer of 1974, war flared, and when it halted a month later, Varosha's Greek Cypriots saw their grand investment end up on the Turkish side of the Green Line. They and all Varosha's residents had to flee south, to the Greek side of the island.

About the size of Connecticut, mountainous Cyprus floats in a placid aquamarine sea ringed by several countries whose genetically intertwined peoples often detest each other. Ethnic Greeks arrived on Cyprus some 4,000 years ago, and subsequently lived under a parade of conquering Assyrians, Phoenicians, Persians, Romans, Arabs, Byzantines, English crusaders, French, and Venetians. The year 1570 brought yet another conqueror, the Ottoman Empire. With it came Turkish settlers, who by the 20th century would comprise slightly less than one-fifth of the island's population.

After World War I finished off the Ottomans, Cyprus ended up as a British colony. The island's Greeks, Orthodox Christians who had periodically revolted against the Ottoman Turks, weren't thrilled to have British rulers instead, and clamored for unification with Greece. The Turkish Cypriot Muslim minority protested. Tensions boiled for decades and erupted viciously several times during the 1950s. A 1960 compromise resulted in the independent Republic of Cyprus, with power shared between Greeks and Turks.

Ethnic hatred, however, had by then become a habit: Greeks massacred entire Turkish families, and Turks ferociously avenged them. A military takeover in Greece detonated a coup on the island, midwifed by the American CIA in honor of Greece's new anticommunist rulers. That prompted Turkey in July 1974 to send troops to protect Turkish Cypriots from being annexed by Greece. During the ensuing brief war, each side was accused of inflicting atrocities on the other's civilians. When the Greeks placed anti-aircraft guns atop a high-risc in the seaside resort of

Varosha, Turkish bombers attacked in American-made jets, and Varosha's Greeks ran for their lives.

Allan Cavinder, a British electrical engineer, had arrived on the island two years earlier, in 1972. He had been taking assignments with a London firm throughout the Middle East, and when he saw Cyprus, he decided to stay. Except for torrid July and August, the island's weather was mostly mild and spotless. He settled on the northern shore, below mountains where yellow limestone villages lived off the harvests of olive and carob trees, which they exported from an inlet harbor at his town, Kyrenia.

When the war began, he decided to wait it out, figuring correctly that there would be demand for his expertise when it ended. He wouldn't have predicted the call from the hotel, however. After the Greeks abandoned Varosha, the Turkish Cypriots, rather than let squatters colonize it, decided that the fancy resort would be more valuable as a bargaining chip when negotiations for a permanent reconciliation got underway. So they built a chain-link fence around it, strung barbed wire across the beach, stationed Turkish soldiers to guard it, and posted signs warning everyone else away.

After two years, however, an old Ottoman foundation that owned property that included the northernmost Varosha hotel requested permission to refurbish and reopen it. It was a sensible idea, Cavinder could see. The four-story hotel, to be renamed the Palm Beach, sat far enough back on a shoreline bend that its terrace and beachfront remained sunny through the afternoon. The hotel tower next-door, which had briefly held a Greek machine-gun placement, had collapsed during the Turkish bombing raid, but aside from its rubble everything else Allan Cavinder found when he first entered the zone seemed intact.

Eerily so: he was struck by how quickly humans had abandoned it. The hotel registry was still open to August 1974, when business had suddenly halted. Room keys lay where they'd been tossed on the front desk. Windows facing the sea had been left open, and blowing sand had formed small dunes in the lobby. Flowers had dried in vases; Turkish coffee demitasses and breakfast dishes licked clean by mice were still in place on the table linens.

His task was to bring the air-conditioning system back into service.

Abandoned hotel, Varosha, Cyprus.

PHOTO BY PETER YATES—IMAGE REPRODUCTION BY 'SOLE STUDIO.

However, this routine job was proving difficult. The southern, Greek portion of the island had UN recognition as the legitimate Cyprus government, but a separate Turkish state in the north was recognized only by Turkey. With no access to spare parts, an arrangement was made with Turkish troops guarding Varosha to allow Cavinder to quietly cannibalize whatever he needed from the other vacant hotels.

He wandered through the deserted town. About 20,000 people had lived or worked in Varosha. Asphalt and pavement had cracked; he wasn't surprised to see weeds growing in the deserted streets, but hadn't expected to see trees already. Australian wattles, a fast-growing acacia species used by hotels for landscaping, were popping out midstreet, some nearly three feet high. Creepers from ornamental succulents snaked out of hotel gardens, crossing roads and climbing tree trunks. Shops still displayed souvenirs and tanning lotion; a Toyota dealership was showing

1974 Corollas and Celicas. Concussions from Turkish air force bombs, Cavinder saw, had exploded plate-glass store windows. Boutique mannequins were half-clothed, their imported fabrics flapping in tattered strips, the dress racks behind them full but deeply dust coated. The canvas of baby prams was likewise torn—he hadn't expected to see so many left behind. And bicycles.

The honeycombed facades of empty hotels, 10 stories of shattered sliding glass doors opening to seaview balconies now exposed to the elements, had become giant pigeon roosts. Pigeon droppings coated everything. Carob rats nested in hotel rooms, living off Yaffa oranges and lemons from former citrus groves that had been absorbed into Varosha's landscaping. The bell towers of Greek churches were spattered with the blood and feces of hanging bats.

Sheets of sand blew across avenues and covered floors. What surprised him at first was the general absence of smell, except for a mysterious stench that emanated from hotel swimming pools, most of which were inexplicably drained yet reeked as though filled with cadavers. Around them, the upturned tables and chairs, torn beach umbrellas, and glasses knocked on their sides all spoke of some revelry gone terribly wrong. Cleaning all this up was going to be expensive.

For six months, as he dismantled and rescued air conditioners, industrial washers and dryers, and entire kitchens full of ovens, grills, refrigerators, and freezers, silence pounded at him. It actually hurt his ears, he told his wife. During the year before the war, he'd worked at a British naval base south of town, and would often leave her at a hotel to enjoy a day at the beach. When he picked her up afterward, a dance band would be playing for the German and British tourists. Now, no bands, just the incessant kneading of the sea that no longer soothed. The wind sighing through open windows became a whine. The cooing of pigeons grew deafening. The sheer absence of human voices bouncing off walls was unnerving. He kept listening for Turkish soldiers, who were under instructions to shoot looters. He wasn't certain how many assigned to patrol knew that he was there legally, or would give him a chance to prove it.

It turned out not to be a problem. He seldom saw any guards. He understood why they would avoid entering such a tomb.

———

By the time Metin Münir saw Varosha, four years after Allan Cavinder's reclamation job ended, roofs had collapsed and trees were growing straight out of houses. Münir, one of Turkey's best-known newspaper columnists, is a Turkish Cypriot who went to Istanbul to study, came home to fight when the troubles began, then returned to Turkey when the troubles kept going on, and on. In 1980, he was the first journalist allowed to enter Varosha for a few hours.

The first thing he noticed was shredded laundry still hanging from clotheslines. What struck him most, though, wasn't the absence of life but its vibrant presence. With the humans who built Varosha gone, nature was intently recouping it. Varosha, merely 60 miles from Syria and Lebanon, is too balmy for a freeze-thaw cycle, but its pavement was tossed asunder anyway. The wrecking crews weren't just trees, Münir marveled, but also flowers. Tiny seeds of wild Cyprus cyclamen had wedged into cracks, germinated, and heaved aside entire slabs of cement. Streets now rippled with white cyclamen combs and their pretty, variegated leaves.

"You understand," Münir wrote his readers back in Turkey, "just what the Taoists mean when say they say that soft is stronger than hard."

Two more decades passed. The millennium turned, and kept going. Once, Turkish Cypriots had bet that Varosha, too valuable to lose, would force the Greeks to the bargaining table. Neither side had dreamed that, 30-plus years later, the Turkish Republic of Northern Cyprus would still exist, severed not only from the Greek Republic of Cyprus but from the world, still a pariah nation to all except Turkey. Even the UN Peacekeeping Force was exactly where it was in 1974, still listlessly patrolling the Green Line, occasionally waxing a pair of still-impounded, still-new 1974 Toyotas.

Nothing has changed except Varosha, which is entering advanced stages of decay. Its encircling fence and barbed wire are now uniformly rusted, but there is nothing left to protect but ghosts. An occasional Coca Cola sign and broadsides posting nightclubs' cover charges hang on doorways that haven't seen customers in more than three decades, and now never will again. Casement windows have flapped and stayed open, their pocked frames empty of glass. Fallen limestone facing lies in pieces. Hunks of wall have dropped from buildings to reveal empty rooms, their furniture long ago somehow spirited away. Paint has dulled; the underlying plaster, where it remains, has yellowed to muted patinas. Where it doesn't, brick-shaped gaps show where mortar has already dissolved.

Other than the back-and-forth of pigeons, all that moves is the creaky rotor of one last functioning windmill. Hotels—mute and windowless, some with balconies that have fallen, precipitating cascades of damage below—still line the riviera that once aspired to be Cannes or Acapulco. At this point, all parties agree, none is salvageable. Nothing is. To someday once again lure tourists, Varosha will have to be bulldozed and begun anew.

In the meantime, nature continues its reclamation project. Feral geraniums and philodendrons emerge from missing roofs and pour down exterior walls. Flame trees, chinaberries, and thickets of hibiscus, oleander, and passion lilac sprout from nooks where indoors and outdoors now blend. Houses disappear under magenta mounds of bougainvillaea. Lizards and whip snakes skitter through stands of wild asparagus, prickly pear, and six-foot grasses. A spreading ground cover of lemon grass sweetens the air. At night, the darkened beachfront, free of moonlight bathers, crawls with nesting loggerhead and green sea turtles.

THE ISLAND OF Cyprus is shaped like a skillet, with its long handle extended toward the Syrian shore. Its pan is gridded with two mountain ranges oriented east-west and divided by a wide central basin—and by the Green Line, with one sierra on each side. The mountains were once covered with Aleppo and Corsican pines, oaks, and cedars. A cypress and juniper forest filled the entire central plain between the two ranges. Olive, almond, and carob trees grew on the arid seaward slopes. At the end of the Pleistocene, dwarf elephants the size of cows and pygmy hippopotamuses no bigger than farmyard swine roamed among these trees. Since Cyprus originally rose from the sea, unconnected to the three continents that surround it, both species apparently arrived by swimming. They were followed by humans about 10,000 years ago. At least one archaeological site suggests that the last pygmy hippo was killed and cooked by *Homo sapiens* hunters.

The trees of Cyprus were prized by Assyrian, Phoenician, and Roman boat builders; during the Crusades, most of them disappeared into the warships of Richard the Lionhearted. By then, the goat population was so large that the plains remained treeless. During the 20th century, plantations of umbrella pines were introduced to try to resuscitate former springs. However, in 1995, following a long drought, nearly all of them and the remaining native forests in the northern mountains exploded in a lightning inferno.

Journalist Metin Münir was too grieved to return again from Istanbul to face his native island in ashes, until a Turkish Cypriot horticulturist, Hikmet Uluçan, convinced him he needed to see what was happening. Once again, Münir found that flowers were renewing a Cyprus landscape: the burnt hillsides were blanketed with crimson poppies. Some poppy seeds, Uluçan told him, live 1,000 years or more, waiting for fire to clear trees away so they can bloom.

In the village of Lapta, high above the northern coastline, Hikmet Uluçan grows figs, cyclamens, cacti, and grapes, and reverently tends the oldest weeping mulberry in all Cyprus. His moustache, Vandyke beard, and remaining tufts of hair have whitened since he was forced to leave the South as a young man, where his father had a vineyard and raised sheep, almonds, olives, and lemons. Until the senseless feud that tore apart his island, 20 generations of Greeks and Turks had shared their valley. Then neighbors were clubbed to death. They found the smashed body of an old Turkish woman who had been grazing her goat, the bleating animal still tied to her wrist. It was barbaric, but Turks were also slaying Greeks. Murderous, mutual loathing between tribes was no more explicable, or complicated, than the genocidal urges of chimpanzees—a fact of nature that we humans, vainly or disingenuously, pretend our codes of civilization transcend.

From his garden, Hikmet can see down to the harbor at Kyrenia, guarded by a seventh-century Byzantine castle built atop Roman fortifications that preceded it. Crusaders and Venetians subsequently took it; then came Ottomans, then British, and now Turks were having a turn again. Today a museum, the castle holds one of the world's rarest relics, a complete Greek merchant ship discovered in 1965, scuttled a mile off Kyrenia. When it went down, its hold was filled with millstones and hundreds of ceramic urns containing wine, olives, and almonds. Its cargo was heavy enough to mire it where currents buried it in mud. Carbon dating of the almonds it carried, likely picked in Cyprus only days earlier, shows that it sank about 2,300 years ago.

Shielded from oxygen, the ship's Aleppo pine hull and timbers remained intact, although they had to be injected with polyethylene resins to keep from disintegrating once exposed to air. The boat builders had used nails of copper, also once plentiful on Cyprus, also impervious to rust. Equally well preserved are lead fishing weights and the ceramic urns whose varied styles reveal the Aegean ports of their origin.

The 10-foot-thick walls and curved towers of the castle where the ship is now displayed are of limestone quarried from the surrounding cliffs, bearing tiny fossils deposited when Cyprus was beneath the Mediterranean. Since the island was divided, however, the castle and the fine old stone carob warehouses of Kyrenia's waterfront have all but disappeared behind unlovely infestations of casino hotels—gambling and loose currency laws being among the limited economic options open to a pariah nation.

Hikmet Uluçan drives east along Cyprus's north coast, passing three more castles of native limestone rising from the jagged mountains that parallel the narrow road. Along the headlands and promontories overlooking the topaz Mediterranean are remains of stone villages, some of them 6,000 years old. Until recently, their terraces, half-buried walls, and jetties were also visible. Since 2003, however, yet another foreign incursion has assaulted the island's profile. "The only consolation," mourns Uluçan, "is that this one can't last."

Not crusaders, this time, but elderly British seeking the warmest retirement a middle-class pension can buy, led by a frenzy of developers who discovered in the quasi-country of Northern Cyprus the last cheap, untouched seafront property left anywhere north of Libya, with pliable zoning codes to match. Suddenly, bulldozers were scattering 500-year-old olive trees to scrape roads across hillsides. Waves of red-tiled roofs soon oscillated across the landscape, atop floor plans cloned repeatedly in poured concrete. Upon a tsunami of cash payoffs, property agents surfed to shore astride English-language billboards that appended terms like "Estates," "Hillside Villas," "Seaside Villas," and "Luxury Villas" to ancient Mediterranean place-names.

Prices from £40,000 to £100,000 ($75,000 to $185,000 U.S.) touched off a land rush that overwhelmed trifles such as title disputes by Greek Cypriots who still claimed to own much of the land. A Northern Cyprus environmental-protection trust vainly protested a new golf course by reminding people that they now had to import water from Turkey in giant vinyl bags; that municipal garbage tips were full; that the total lack of sewage treatment meant five times as much effluent would pour into the transparent sea.

Each month more steam shovels gobble coastline like famished brontosaurs, spitting out olive and carob trees along a widening blacktop now 30 miles east of Kyrenia, with no sign of stopping. The English language marches down the shore, dragging embarrassing architecture with it, one sign after another announcing the latest subdivision with a trust-inspiring British name, even as the seaside villas grow trashier: concrete painted, not stuccoed; fake-ceramic roof tiles made of tacky polymer; cornices and windows trimmed with faux stenciled stonework. When Hikmet Uluçan sees a pile of traditional yellow tiles lying in front of naked town house frames awaiting walls, he realizes that someone is ripping stone facing from local bridges and selling it to contractors.

Something about these limestone squares lying at the base of skeletal buildings looks familiar. After a while, he figures it out. "It is like Varosha." The half-finished buildings going up, surrounded by construction rubble, exactly recall the half-ruins of Varosha coming down.

But if anything, quality has sunk even further. Each billboard touting Northern Cyprus's sunny new dream homes includes, near the bottom, notification of the construction guarantee: 10 years. Given rumors of developers not bothering to wash the sea salt from the beach sand they mine for concrete, 10 years may be all they get.

Beyond the new golf course, the road finally narrows again. Past one-lane bridges stripped of their limestone ornaments, and a small canyon filled with myrtle and pink orchids, it enters the Karpaz Peninsula, the long tendril that reaches east toward the Levant. Along it are empty Greek churches, gutted but going nowhere, testaments to the tenacity of stone architecture. Stone structures were among the first things that distinguished sedentary humans from nomadic hunter-gatherers, whose temporary mud-and-wattle huts were no more permanent than the season's grass. Stone buildings will be among the last to disappear when we're gone. As the fleeting materials of modern construction decompose, the world will retrace our steps back to the Stone Age as it gradually erodes away all memory of us.

As the road follows the peninsula, the landscape gets biblical, with old walls turning back to hills as gravity tugs on the underlying clay. The island ends in sand dunes covered by salt scrub and pistachio trees. The beach is smoothed by the belly tracks of mama sea turtles.

A small limestone hill is topped by a lone, branching umbrella pine. Shadows on the rock face turn out to be caves. Closer, the soft parabola of a low-arched portal reveals that it's been carved. At this windblown land's end, less than 40 miles across the water to Turkey and only 20 miles more to Syria, the Stone Age began in Cyprus. Humans arrived around the same time that the oldest known building on Earth, a stone tower, was rising in the world's oldest city that is still inhabited, Jericho. However primitive this Cyprus dwelling is by comparison, it represented a momentous step, albeit one taken some 40,000 years earlier by Southeast Asians who reached Australia—seafarers venturing beyond the horizon, out of sight of the shore, and finding another one waiting.

The cave is shallow, perhaps 20 feet deep, and surprisingly warm. A charcoal-smudged hearth, two benches, and sleeping niches were fashioned by cutting into the sedimentary walls. A second room, smaller than the first, is almost square, with a squared doorway arch.

Remains of *Australopithecus* in South Africa suggest that we were cavemen at least 1 million years ago. In a river bluff grotto at Chauvet-Pont-d'Arc in France, Cro-Magnons not only occupied caves 32,000 years ago, but also turned them into our first art galleries, depicting the European megafauna they sought, or whose strength they wished to channel.

There are no such artifacts here: these first inhabitants of Cyprus were struggling pioneers, their time for aesthetic reflection still ahead. But their bones are buried beneath the floor. Long after all our buildings and what's left of the tower at Jericho are reduced to sand and soil, caves where we took shelter and first learned the notion of walls—including that they begged for art—will remain. In a world without us, they await the next occupant.

ᵙᘒ

What Lasts

1. Earth and Sky Tremors

I**T IS HARD** to see exactly what holds up the enormous round dome of Istanbul's formerly Orthodox Christian, marble and mosaic-encrusted church of Hagia Sophia. More than 100 feet across, it is slightly smaller than the dome that crowns Rome's Pantheon, but considerably higher. An inspired stroke of design divides its weight through a colonnade of arched windows at its base, making it appear to float. To gaze straight up at it, a gilded sky hovering 185 feet overhead, with no easy sense of why it stays aloft, leaves a beholder half-believing in miracles, and half-dizzy.

Over a thousand years, the dome's weight has been further distributed among so many redoubled interior walls, additional half-domes, flying buttresses, pendentives, and massive corner piers that Turkish civil engineer Mete Sözen believes that not even a major earthquake would easily shake it loose. That is exactly what happened to its first dome, which fell just 20 years after it was completed in AD 537. That mishap led to all the subsequent reinforcement; even so, quakes severely damaged the church (which became a mosque in 1453) twice more until Mimar Sinan, the greatest architect of the Ottoman Empire, restored it in the 16th century. The delicate minarets the Ottomans added to its exterior will one day topple, but even in a world without people, meaning no masons to periodically repoint the Hagia Sophia's mortar, Sözen expects much of it and other great ancient masonry edifices of Istanbul to last well into future geologic time.

Which is more than he can say, unfortunately, for the rest of the city of his birth. Not that it's quite the same city. Through history, Istanbul, née Constantinople, née Byzantium before that, has changed hands so many times that it's hard to imagine what could fundamentally alter it, let alone destroy it. But Mete Sözen is convinced that the former has already happened and the latter is imminent, whether humans stick around or not. The only difference in a world without people is that no one would be left to try to pick up Istanbul's pieces.

When Dr. Sözen, who holds a chair in structural engineering at Indiana's Purdue University, first left Turkey in 1952 for graduate studies in the United States, Istanbul had 1 million people. Half a century later, it has 15 million. He describes that as a far greater paradigm shift than its previous transformations from Delphic to Roman to Byzantine Orthodox to Crusader Catholic and, finally, to Muslim—in all its Ottoman and Turkish Republican strains.

Dr. Sözen sees this difference through an engineer's eyes. Whereas all the previous conquering cultures erected fabulous monuments to themselves like the Hagia Sophia and the nearby ethereal Blue Mosque, the architectural expression of today's hordes is manifest in more than 1 million multi-story buildings jammed into Istanbul's narrow streets—buildings that he says are doomed to abbreviated life spans. In 2005, Sözen and a team he assembled of international architectural and seismic experts warned the Turkish government that within 30 years, the North Anatolian Fault that runs just east of the city will slip again. When it does, at least 50,000 apartment buildings will fall.

He's still awaiting a response, although he doubts that anyone can imagine where to begin to stave off what his expertise deems inevitable. In September 1985, the U.S. government rushed Sözen to Mexico City to analyze how its embassy had weathered an 8.1-magnitude earthquake that collapsed nearly 1,000 buildings. The highly reinforced embassy, which he had examined a year earlier, was intact. Up and down Avenida Reforma and adjacent streets, however, many high-rise offices, apartments, and hotels had disintegrated.

It was one of the worst quakes in Latin American history. "But it was mostly confined to downtown. What occurred in Mexico City is a flake of what will happen to Istanbul."

One thing that the two disasters, past and future, have in common is that nearly all the buildings that crumbled or will crumble were built after

World War II. Turkey stayed out of that war, but its economy took the same beating as every other country's. As industries recovered in the post-war European boom, thousands of peasants migrated to cities everywhere seeking jobs. Both the European and Asian sides of the Bosphorus Strait, which Istanbul straddles, filled with six- and seven-story housing of rein-forced concrete.

"But the quality of the concrete," Mete Sözen told the Turkish gov-ernment, "is 1/10 of what you'd find in, say, Chicago. Strength and qual-ity of concrete depend on the amount of cement used."

Back then, the problem was economics and availability. But as Istan-bul's population grew, the problem did, too, with more floors added to accommodate more humans. "The success of a concrete or masonry build-ing," Sözen explained, "depends on how much you have to support above the first level. The more floors, the heavier the building." The danger comes when residential stories are stacked on top of structures whose ground floors are used for shops or restaurants. Most are open commercial spaces that lack internal columns or load-bearing walls because they were never intended to support more than one story.

Complicating matters further, floors added as afterthoughts rarely align in adjacent buildings, placing uneven stress on shared walls. Worse still, Sözen said, is when space is left at the top of a wall for ventilation, or to save material. When a building sways during an earthquake, exposed columns in partial walls shear off. In Turkey, hundreds of schools have just such a design. Wherever air-conditioning is unaffordable in the trop-ics, from the Caribbean to Latin America to India to Indonesia, these ex-tra spaces are especially common as a way to bleed off heat and invite breezes. In the developed world, the identical weakness is often found in structures without climate control, such as parking garages.

In a 21st century where more than half the human race lives in cities and where most people are poor, cheap variations on the theme of rein-forced concrete are repeated daily: planet-wide piles of low bids that will come crashing down in a posthuman world, and do so even faster if the city is near a fault line. When an earthquake strikes Istanbul, its narrow, winding streets will clog so totally with the rubble of thousands of wrecked buildings, Sözen estimates, that much of the city will simply have to close down for 30 years before the massive destruction can be cleared away.

Assuming there is anybody to do the clearing. If not, and if Istanbul remains a city where snow regularly falls in the winter, then freeze-thaw cycles will have plenty of earthquake detritus to reduce to sand and soil above the cobbles and pavement. Every earthquake causes fires; in the absence of response crews, the grand old wooden Ottoman mansions along the Bosphorus will contribute the ash of long-extinct cedars to the formation of new soil.

Although mosque domes, like the Hagia Sophia's, will initially survive, the shaking will have loosened their masonry, and freeze-thaw will work at their mortar until bricks and stones start to fall. Eventually, as in 4,000-year-old Troy 175 miles down Turkey's Aegean coast, only Istanbul's roofless temple walls will remain—still standing, but buried.

2. Terra Firma

Should Istanbul exist long enough to complete its planned subway system—including a line under the Bosphorus that would link Europe and Asia—since its tracks will cross no fault line, it will probably remain intact, albeit forgotten, long after the city on the surface is gone. (Subways whose tunnels *do* encounter geologic faults, however, such as the San Francisco Bay Area's BART and New York City's MTA, may face another fate.) In the Turkish capital of Ankara, the subway system's central nerve core broadens into an extensive underground shopping district with mosaic walls, acoustic ceilings, electronic billboard screens, and arcades of stores—an orderly underworld compared to the cacophony of the streets above.

Ankara's sub-surface shops; Moscow's subway, with its deep train tunnels and chandelier-lit, museum-like underground stations, renowned as some of the most elegant spots in the city; Montreal's subterranean village of shops, malls, offices, apartments, and labyrinthine passages that reflect the city in miniature and link its old-fashioned surface structures—all these underground creations stand the best chance of any man-made edifices of lasting into whatever hereafter lies beyond human existence on Earth. Although seepage and surface cave-ins will eventually reach them, buildings still exposed to the elements will go well before structures that were born already buried.

These won't be the oldest, however. Three hours south of Ankara is a region of central Turkey whose name, Cappadocia, ostensibly means "Land of Fine Horses." But that has to be a mistake: the result, possibly, of some garbled pronunciation of a more fitting description in some ancient tongue, because not even winged horses could steal the spotlight from this landscape—or from what lies beneath it.

In 1963, a fresco now thought to be the oldest landscape painting on Earth was discovered in Turkey by University of London archaeologist James Mellaart. Between 8,000 and 9,000 years old, it is also the oldest known work rendered on a surface constructed by humans: in this case, a mud-brick plastered wall. Overtly two-dimensional, the eight-foot mural is a flattened image of an erupting, double-coned volcano. Out of context, its components make little sense: The volcano itself, painted with ochre pigments worked into wet lime plaster, could be mistaken for a bladder, or even two disembodied breasts—in this case, the teats of a female leopard, as they are curiously flecked with black spots. The volcano also seems to be perched directly atop a pile of boxes.

From the vantage point of where it was discovered, however, its meaning is unmistakable. The double-barreled volcano's shape matches the silhouette of 10,700-foot Hasan Dağ 40 miles to the east, a long swoop of a mountain that hangs over central Turkey's high Konya plain. Together, the boxes form a primitive town plat of what many scholars consider the world's first city, Çatal Höyük, which is twice as old as the pyramids of Egypt—and, with a population around 10,000, was far bigger than its contemporary, Jericho.

All that remained of it when Mellaart began digging was a low mound rising above wheat and barley fields. The first things he found were hundreds of obsidian points, which may explain the black spots, as the Hasan Dağ volcano was the source for that substance. But for reasons unknown, Çatal Höyük had been abandoned. The mud-brick walls of its house-sized boxes had fallen in on themselves, and erosion smoothed the rectangles of its skyline into a gentle parabola. Another 9,000 years, and the parabola should be long flattened.

On Hasan Dağ's opposite slope, however, something quite different happened. What is now called Cappadocia began as a lake. During millions of years of frequent volcanic eruptions, its bowl filled with layers of ash that kept piling on, hundreds of feet deep. When the cauldron finally cooled, these congealed into tuff, a rock with remarkable properties.

A huge, final burst 2 million years ago unrolled a mantle of lava that left a thin crust of basalt atop 10,000 square miles of powdery gray tuff. When it hardened, so did the climate. Rain, wind, and snow set to work, with freeze-thaw cycles cracking and splitting the basalt pavement, and moisture seeping in to dissolve the tuff below. As it eroded, in places the ground collapsed. Left standing were hundreds of pale, slender pinnacles, each mushroom-capped with a hood of darker basalt.

Tourism promoters call them fairy towers, a plausible descriptor but not necessarily the first one that comes to mind. The magical version prevails, however, because the surrounding tuff hills have invited not just wind and water to sculpt them, but also the hands of imaginative humans. Cappadocia's towns have not been built so much on the land, as in it.

Tuff is soft enough that a determined prisoner here could scoop his way out of a dungeon with a spoon. When exposed to air, however, it hardens, forming a smooth, stucco-like shell. By 700 BC, humans with iron tools were burrowing into Cappadocia's escarpments, and even hollowing out fairy towers. Like a prairie dog village tipped on its side, every rock face was soon riddled with holes—some big enough for a pigeon, or a person, or a three-floor hotel.

The pigeonholes—hundreds of thousands of arched niches hollowed into valley walls and pinnacles—were intended to attract rock doves for exactly the same reason humans in modern cities try to chase their urban cousins away: their copious droppings. So prized was pigeon guano, used here to nourish grapes, potatoes, and famously sweet apricots, that the carved exteriors of many dovecotes bear flourishes as ornate as those found on Cappadocia's cave churches. This architectural homage to a feathered fellow creature continued until artificial fertilizers reached here in the 1950s. Since then, Cappadocians no longer build them. (Nor do they now build churches. Before the Ottomans converted Turkey to Islam, more than 700 were cut into Cappadocia's plateaus and mountainsides.)

Much of today's most expensive real estate here consists of luxury homes carved into tuff, with bas-relief exteriors as pretentious as the facades of mansions anywhere, and with mountainside views to match.

Former churches have been recast into mosques; the muezzin call to evening devotion, resounding among Cappadocia's slick tuff walls and spires, is like a congregation of mountains praying.

One distant day, these man-made caves—and even natural ones, of stone much harder than volcanic tuff—will wear away. In Cappadocia, however, the stamp of humanity's passage will linger beyond our other traces, because here humans have not only ensconced themselves in plateau walls, but also beneath the plains. Deep beneath. Should the Earth's poles shift and sheet glaciers one day muscle their way across central Turkey, flattening whatever man-made structures still stand in their way, here they will only scratch our surface.

No one knows how many underground cities lie beneath Cappadocia. Eight have been discovered, and many smaller villages, but there are doubtless more. The biggest, Derinkuyu, wasn't discovered until 1965, when a resident cleaning a back room of his cave house broke through a wall and discovered behind it a room that he'd never seen, which led to still another, and another. Eventually, spelunking archaeologists found a maze of connecting chambers that descended at least 18 stories and 280 feet beneath the surface, ample enough to hold 30,000 people—and much remains to be excavated. One tunnel, wide enough for three people walking abreast, connects to another underground town six miles away. Other passages suggest that at one time all of Cappadocia, above and below the ground, was linked by a hidden network. Many still use the tunnels of this ancient subway as cellar storerooms.

Unlike a river canyon, the earliest segments here are nearest the surface. Some believe the first builders were the Hittites of biblical times, who burrowed underground to hide from marauding Phrygians. Murat Ertuğrul Gülyaz, an archaeologist at Cappadocia's Nevşehir Museum, agrees that Hittites lived here, but doubts they were the first.

Gülyaz, a proud native with a moustache thick as a fine Turkish rug, worked on the excavation of Aşikli Höyük, a small Cappadocian mound containing the remains of a settlement even older than Çatal Höyük. Among the relics there were 10,000-year-old stone axes and obsidian tools capable of cutting tuff. "The underground cities are prehistoric," he declares. That, he says, explains the crudeness of the upper chambers,

Underground city, Derinkuyu, Cappadocia, Turkey.
PHOTO BY MURAT ERTUĞRUL GÜLYAZ.

compared to the precision of the rectangular floors below. "Later, every-one who appeared kept going deeper."

It's as if they couldn't stop, one conquering culture after another real-izing the benefit of a hidden, sub-surface world. The underground cities were lit by torches, or often, Gülyaz discovered, by linseed oil lamps, which also gave enough heat to keep temperatures pleasant. Temperature was probably what first inspired humans to dig them, for winter shelter. But as successive waves of Hittites, Assyrians, Romans, Persians, Byzan-tines, Seljuk Turks, and Christians discovered these dens and warrens, they widened and deepened them for one principal reason: defense. The last two even expanded the original upper chambers enough to stable their horses underground.

The smell of tuff that permeates Cappadocia—cool, clayish, with a men-thol tang—intensifies below. Its versatile nature allowed niches to be scooped where lamplight was needed, yet tuff is strong enough that

Turkey considered using these nether cities as bomb shelters had the 1990 Persian Gulf War spread.

In the underground city of Derinkuyu, the floor below the stables held fodder bins for livestock. Next down was a communal kitchen, with earthen ovens placed below holes in nine-foot ceilings that, via offset rock tubes, channeled smoke to chimneys two kilometers away, so that enemies wouldn't know where they were. For the same reason, ventilation shafts were also engineered on the skew.

Copious storage space and thousands of earthenware jars and urns suggest that thousands of people spent months down here without seeing the sun. Through vertical communication shafts, it was possible to speak to another person on any level. Underground wells provided their water; underground drains prevented flooding. Some water was routed through tuff conduits to underground wineries and breweries, equipped with tuff fermentation vats and basalt grinding wheels.

These beverages were probably essential for calming the claustrophobia induced by passing between levels via staircases so intentionally low, tight, and serpentine that any invaders had to proceed slowly, bent over, and in single file. Emerging one by one, they would be easily slain—if they got that far. Stairways and ramps had landings every 10 meters, with Stone Age pocket doors—half-ton, floor-to-ceiling stone wheels—that could be rolled in place to seal a passage. Trapped between a pair of these, intruders would soon notice that holes overhead weren't air shafts, but pipes for bathing them with hot oil.

Another three floors below this underworld fortress, a room with a vaulted ceiling and benches facing a stone lectern was a school. Farther below were multiple levels of living quarters, strung along underground streets that branched and intersected for several square kilometers. They included double alcoves for adults with children, and even playrooms featuring pitch-black tunnels that returned to the same spot.

And farther: eight levels down in Derinkuyu, two large, high-ceilinged spaces join in a cruciform. Although, due to constant humidity, no frescoes or paintings remain, in this church, seventh-century Christians who emigrated from Antioch and Palestine would have prayed and hidden from Arab invaders.

Below it is a tiny, cube-shaped room. It was a temporary tomb, where the dead could be kept until danger passed. As Derinkuyu and the other underground cities passed from hand to hand and civilization to civilization, their citizens always returned to the surface, to bury their own in the soil where food grew under sun and rainfall.

The surface was where they were bred to live and die, but one day when we are long gone, it is the underground cities they built for protection that will defend humanity's memory, bearing final—albeit hidden—witness to the fact that, once, we were here.

❧

CHAPTER 9

Polymers Are Forever

THE PORT OF Plymouth in southwestern England is no longer listed among the scenic towns of the British Isles, although prior to World War II it would have qualified. During six nights of March and April 1941, Nazi bombs destroyed 75,000 buildings in what is remembered as the Plymouth Blitz. When the annihilated city center was rebuilt, a modern concrete grid was superimposed on Plymouth's crooked cobbled lanes, burying its medieval past in memory.

But the main history of Plymouth lies at its edge, in the natural harbor formed at the confluence of two rivers, the Plym and the Tamar, where they join the English Channel and the Atlantic Ocean. This is the Plymouth from which the Pilgrims departed; they named their American landfall across the sea in its honor. All three of Captain Cook's Pacific expeditions began here, as did Sir Francis Drake's circumnavigation of the globe. And, on December 27, 1831, H.M.S. *Beagle* set sail from Plymouth Harbor, with 22-year-old Charles Darwin aboard.

University of Plymouth marine biologist Richard Thompson spends a lot of time pacing Plymouth's historic edge. He especially goes in winter, when the beaches along the harbor's estuaries are empty—a tall man in jeans, boots, blue windbreaker, and zippered fleece sweater, his bald pate hatless, his long fingers gloveless as he bends to probe the sand. Thompson's doctoral study was on slimy stuff that mollusks such as limpets and winkles like to eat: diatoms, cyanobacteria, algae, and tiny plants that cling to seaweed. What he's now known for, however, has less to do with marine life than with the growing presence of things in the ocean that have never been alive at all.

Although he didn't realize it at the time, what has dominated his life's work began when he was still an undergraduate in the 1980s, spending autumn weekends organizing the Liverpool contingent of Great Britain's national beach cleanup. In his final year, he had 170 teammates amassing metric tons of rubbish along 85 miles of shoreline. Apart from items that apparently had dropped from boats, such as Greek salt boxes and Italian oil cruets, from the labels he could see that most of the debris was blowing east from Ireland. In turn, Sweden's shores were the receptacles for trash from England. Any packaging that trapped enough air to protrude from the water seemed to obey the wind currents, which in these latitudes are easterly.

Smaller, lower-profile fragments, however, were apparently controlled by currents in the water. Each year, as he compiled the team's annual reports, Thompson noticed more and more garbage that was smaller and smaller amid the usual bottles and automobile tires. He and another student began collecting sand samples along beach strand lines. They sieved the tiniest particles of whatever appeared unnatural, and tried to identify them under a microscope. This proved tricky: their subjects were usually too small to allow them to pinpoint the bottles, toys, or appliances from which they sprang.

He continued working the annual cleanup during graduate studies at Newcastle. Once he completed his Ph.D. and began teaching at Plymouth, his department acquired a Fourier Transform Infrared Spectrometer, a device that passes a microbeam through a substance, then compares its infrared spectrum to a database of known material. Now he could know what he was looking at, which only deepened his concern.

"Any idea what these are?" Thompson is guiding a visitor along the shore of the Plym River estuary, near where it joins the sea. With a full moonrise just a few hours off, the tide is out nearly 200 meters, exposing a sandy flat scattered with bladderwrack and cockle shells. A breeze skims the tidal pools, shivering rows of reflected hillside housing projects. Thompson bends over the strand line of detritus left by the forward edge of waves lapping the shore, looking for anything recognizable: hunks of nylon rope, syringes, topless plastic food containers, half a ship's float, pebbled remains of polystyrene packaging, and a rainbow of assorted bottle caps. Most plentiful of all are multicolored plastic shafts of cotton ear swabs. But there are also the odd little uniform shapes he challenges people to identify. Amid twigs and seaweed fibers in his fistful of sand are a couple of dozen blue and green plastic cylinders about two millimeters high.

"They're called nurdles. They're the raw materials of plastic production. They melt these down to make all kinds of things." He walks a little farther, then scoops up another handful. It contains more of the same plastic bits: pale blue ones, greens, reds, and tans. Each handful, he calculates, is about 20 percent plastic, and each holds at least 30 pellets.

"You find these things on virtually every beach these days. Obviously they are from some factory."

However, there is no plastic manufacturing anywhere nearby. The pellets have ridden some current over a great distance until they were deposited here—collected and sized by the wind and tide.

In Thompson's laboratory at the University of Plymouth, graduate student Mark Browne unpacks foil-wrapped beach samples that arrive in clear zip-lock bags sent by an international network of colleagues. He transfers these to a glass separating funnel, filled with a concentrated solution of sea salt to float off the plastic particles. He filters out some he thinks he recognizes, such as pieces of the ubiquitous colored ear-swab shafts, to check under the microscope. Anything really unusual goes to the FTIR Spectrometer.

Each takes more than an hour to identify. About one-third turn out to be natural fibers such as seaweed, another third are plastic, and another third are unknown—meaning that they haven't found a match in their polymer database, or that the particle has been in the water so long its color has degraded, or that it's too small for their machine, which analyzes fragments only to 20 microns—slightly thinner than a human hair.

"That means we're underestimating the amount of plastic that we're finding. The true answer is we just don't know how much is out there."

What they do know is that there's much more than ever before. During the early 20th century, Plymouth marine biologist Alistair Hardy developed an apparatus that could be towed behind an Antarctic expedition boat, 10 meters below the surface, to sample krill—an ant-sized, shrimplike invertebrate on which much of the planet's food chain rests. In the

1930s, he modified it to measure even smaller plankton. It employed an impeller to turn a moving band of silk, similar to how a dispenser in a public lavatory moves cloth towels. As the silk passed over an opening, it filtered plankton from water passing through it. Each band of silk had a sampling capacity of 500 nautical miles. Hardy was able to convince English merchant vessels using commercial shipping lanes throughout the North Atlantic to drag his Continuous Plankton Recorder for several decades, amassing a database so valuable he was eventually knighted for his contributions to marine science.

He took so many samples around the British Isles that only every second one was analyzed. Decades later, Richard Thompson realized that the ones that remained stored in a climate-controlled Plymouth warehouse were a time capsule containing a record of growing contamination. He picked two routes out of northern Scotland that had been sampled regularly: one to Iceland, one to the Shetland Islands. His team pored over rolls of silk reeking of chemical preservative, looking for old plastic. There was no reason to examine years prior to World War II, because until then plastic barely existed, except for the Bakelite used in telephones and radios, appliances so durable they had yet to enter the waste chain. Disposable plastic packaging hadn't yet been invented.

By the 1960s, however, they were seeing increasing numbers of increasing kinds of plastic particles. By the 1990s, the samples were flecked with triple the amount of acrylic, polyester, and crumbs of other synthetic polymers than was present three decades earlier. Especially troubling was that Hardy's plankton recorder had trapped all this plastic 10 meters below the surface, suspended in the water. Since plastic mostly floats, that meant they were seeing just a fraction of what was actually there. Not only was the amount of plastic in the ocean increasing, but ever smaller bits of it were appearing—small enough to ride global sea currents.

Thompson's team realized that slow mechanical action—waves and tides that grind against shorelines, turning rocks into beaches—were now doing the same to plastics. The largest, most conspicuous items bobbing in the surf were slowly getting smaller. At the same time, there was no sign that any of the plastic was biodegrading, even when reduced to tiny fragments.

"We imagined it was being ground down smaller and smaller, into a kind of powder. And we realized that smaller and smaller could lead to bigger and bigger problems."

He knew the terrible tales of sea otters choking on polyethylene rings from beer six-packs; of swans and gulls strangled by nylon nets and fishing lines; of a green sea turtle in Hawaii dead with a pocket comb, a foot of nylon rope, and a toy truck wheel lodged in its gut. His personal worst was a study on fulmar carcasses washed ashore on North Sea coastlines. Ninety-five percent had plastic in their stomachs—an average of 44 pieces per bird. A proportional amount in a human being would weigh nearly five pounds.

There was no way of knowing if the plastic had killed them, although it was a safe bet that, in many, chunks of indigestible plastic had blocked their intestines. Thompson reasoned that if larger plastic pieces were breaking down into smaller particles, smaller organisms would likely be consuming them. He devised an aquarium experiment, using bottom-feeding lugworms that live on organic sediments, barnacles that filter organic matter suspended in water, and sand fleas that eat beach detritus. In the experiment, plastic particles and fibers were provided in proportionately bite-size quantities. Each creature promptly ingested them.

When the particles lodged in their intestines, the resulting constipation was terminal. If they were small enough, they passed through the invertebrates' digestive tracts and emerged, seemingly harmlessly, out the other end. Did that mean that plastics were so stable that they weren't toxic? At what point would they start to naturally break down—and when they did, would they release some fearful chemicals that would endanger organisms sometime far in the future?

Richard Thompson didn't know. Nobody did, because plastics haven't been around long enough for us to know how long they'll last or what happens to them. His team had identified nine different kinds in the sea so far, varieties of acrylic, nylon, polyester, polyethylene, polypropylene, and polyvinyl chloride. All he knew was that soon everything alive would be eating them.

"When they get as small as powder, even zooplankton will swallow them."

Two sources of tiny plastic particles hadn't before occurred to Thompson. Plastic bags clog everything from sewer drains to the gullets of sea turtles who mistake them for jellyfish. Increasingly, purportedly biodegradable versions were available. Thompson's team tried them. Most turned out to be just a mixture of cellulose and polymers. After the cellulose starch broke down, thousands of clear, nearly invisible plastic particles remained.

Some bags were advertised to degrade in compost piles as heat generated by decaying organic garbage rises past 100°F. "Maybe they do. But that doesn't happen on a beach, or in salt water." He'd learned that after they tied plastic produce bags to moorings in Plymouth Harbor. "A year later you could still carry groceries in them."

Even more exasperating was what his Ph.D. student Mark Browne discovered while shopping in a pharmacy. Browne pulls open the top drawer of a laboratory cabinet. Inside is a feminine cornucopia of beauty aids: shower massage creams, body scrubs, and hand cleaners. Several are by boutique labels: Neova Body Smoother, SkinCeuticals Body Polish, and DDF Strawberry Almond Body Polish. Others are international name brands: Pond's Fresh Start, a tube of Colgate Icy Blast toothpaste, Neutrogena, Clearasil. Some are available in the United States, others only in the United Kingdom. But all have one thing in common.

"Exfoliants: little granules that massage you as you bathe." He selects a peach-colored tube of St. Ives Apricot Scrub; its label reads, *100% natural exfoliants*. "This stuff is okay. The granules are actually chunks of ground-up jojoba seeds and walnut shells." Other natural brands use grape seeds, apricot hulls, coarse sugar, or sea salt. "The rest of them," he says, with a sweep of his hand, "have all gone to plastic."

On each, listed among the ingredients are "micro-fine polyethylene granules," or "polyethylene micro-spheres," or "polyethylene beads." Or just polyethylene.

"Can you believe it?" Richard Thompson demands of no one in particular, loud enough that faces bent over microscopes rise to look at him. "They're selling plastic meant to go right down the drain, into the sewers, into the rivers, right into the ocean. Bite-size pieces of plastic to be swallowed by little sea creatures."

Plastic bits are also increasingly used to scour paint from boats and aircraft. Thompson shudders. "One wonders where plastic beads laden with paint are disposed. It would be difficult to contain them on a windy day. But even if they're contained, there's no filter in any sewage works for material that small. It's inevitable. They end up in the environment."

He peers into Browne's microscope at a sample from Finland. A lone green fiber, probably from a plant, lies across three bright blue threads that probably aren't. He perches on the countertop, hooking his hiking boots around a lab stool. "Think of it this way. Suppose all human activity ceased tomorrow, and suddenly there's no one to produce plastic anymore. Just from what's already present, given how we see it fragmenting, organisms will be dealing with this stuff indefinitely. Thousands of years, possibly. Or more."

‿❦‿

IN ONE SENSE, plastics have been around for millions of years. Plastics are polymers: simple molecular configurations of carbon and hydrogen atoms that link together repeatedly to form chains. Spiders have been spinning polymer fibers called silk since before the Carboniferous Age, whereupon trees appeared and started making cellulose and lignin, also natural polymers. Cotton and rubber are polymers, and we make the stuff ourselves, too, in the form of collagen that comprises, among other things, our fingernails.

Another natural, moldable polymer that closely fits our idea of plastics is the secretion from an Asian scale beetle that we know as shellac. It was the search for an artificial shellac substitute that one day led chemist Leo Baekeland to mix tarry carbolic acid—phenol—with formaldehyde in his garage in Yonkers, New York. Until then, shellac was the only coating available for electric wires and connections. The moldable result became Bakelite. Baekeland became very wealthy, and the world became a very different place.

Chemists were soon busy cracking long hydrocarbon chain molecules of crude petroleum into smaller ones, and mixing these fractionates to see what variations on Baekeland's first man-made plastic they could produce. Adding chlorine yielded a strong, hardy polymer unlike anything in nature, known today as PVC. Blowing gas into another polymer as it formed created tough, linked bubbles called polystyrene, often known by the brand name Styrofoam. And the continual quest for an artificial silk led to nylon. Sheer nylon stockings revolutionized the apparel industry, and helped to drive acceptance of plastic as a defining achievement of modern life. The intercession of World War II, which diverted most nylon and plastic to the war effort, only made people desire them more.

After 1945, a torrent of products the world had never seen roared into general consumption: acrylic textiles, Plexiglass, polyethylene bottles, polypropylene containers, and "foam rubber" polyurethane toys. Most world-changing of all was transparent packaging, including self-clinging wraps of polyvinyl chloride and polyethylene, which let us see the foods wrapped inside them and kept them preserved longer than ever before.

Within 10 years, the downside to this wonder substance was apparent. *Life Magazine* coined the term "throwaway society," though the idea of tossing trash was hardly new. Humans had done that from the beginning with leftover bones from their hunt and chaff from their harvest, whereupon other organisms took over. When manufactured goods entered the garbage stream, they were at first considered less offensive than smelly organic wastes. Broken bricks and pottery became the fill for the buildings of subsequent generations. Discarded clothing reappeared in secondary markets run by ragmen, or were recycled into new fabric. Defunct machines that accumulated in junkyards could be mined for parts or alchemized into new inventions. Hunks of metal could simply be melted down into something totally different. World War II—at least the Japanese naval and air portion—was literally constructed out of American scrap heaps.

Stanford archaeologist William Rathje, who has made a career of studying garbage in America, finds himself continually disabusing waste-management officials and the general public of what he deems a myth: that plastic is responsible for overflowing landfills across the country. Rathje's decades-long Garbage Project, wherein students weighed and measured weeks' worth of residential waste, reported during the 1980s that, contrary to popular belief, plastic accounts for less than 20 percent by volume of buried wastes, in part because it can be compressed more tightly than other refuse. Although increasingly higher percentages of plastic items have been produced since then, Rathje doesn't expect the proportions to change, because improved manufacturing uses less plastic per soda bottle or disposable wrapper.

The bulk of what's in landfills, he says, is construction debris and paper products. Newspapers, he claims, again belying a common assumption, don't biodegrade when buried away from air and water. "That's why we have 3,000-year-old papyrus scrolls from Egypt. We pull perfectly readable newspapers out of landfills from the 1930s. They'll be down there for 10,000 years."

He agrees, though, that plastic embodies our collective guilt over trashing the environment. Something about plastic feels uneasily permanent. The difference may have to do with what happens outside landfills, where a newspaper gets shredded by wind, cracks in sunlight, and dissolves in rain—if it doesn't burn first.

What happens to plastic, however, is seen most vividly where trash is never collected. Humans have continuously inhabited the Hopi Indian Reservation in northern Arizona since AD 1000—longer than any other site in today's United States. The principal Hopi villages sit atop three mesas with 360° views of the surrounding desert. For centuries, the Hopis simply threw their garbage, consisting of food scraps and broken ceramic, over the sides of the mesas. Coyotes and vultures took care of the food wastes, and the pottery sherds blended back into the ground they came from.

That worked fine until the mid-20th century. Then, the garbage tossed over the side stopped going away. The Hopis were visibly surrounded by a rising pile of a new, nature-proof kind of trash. The only way it disappeared was by being blown across the desert. But it was still there, stuck to sage and mesquite branches, impaled on cactus spines.

South of the Hopi Mesas rise the 12,500-foot San Francisco Peaks, home to Hopi and Navajo gods who dwell among aspens and Douglas firs: holy mountains cloaked in purifying white each winter—except in recent years, because snow now rarely falls. In this age of deepening drought and rising temperatures, ski lift operators who, the Indians claim, defile sacred ground with their clanking machines and lucre, are being sued anew. Their latest desecration is making artificial snow for their ski runs from wastewater, which the Indians liken to bathing the face of God in shit.

East of the San Francisco Peaks are the even taller Rockies; to their west are the Sierra Madres, whose volcanic summits are higher still. Impossible as it is for us to fathom, all these colossal mountains will one day erode to the sea—every boulder, outcrop, saddle, spire, and canyon wall. Every massive uplift will pulverize, their minerals dissolving to keep the oceans salted, the plume of nutrients in their soils nourishing a new marine biological age even as the previous one disappears beneath their sediments.

Long before that, however, these deposits will have been preceded by a substance far lighter and more easily carried seaward than rocks or even grains of silt.

Capt. Charles Moore of Long Beach, California, learned that the day in 1997 when, sailing out of Honolulu, he steered his aluminum-hulled catamaran into a part of the western Pacific he'd always avoided. Sometimes known as the horse latitudes, it is a Texas-sized span of ocean between Hawaii and California rarely plied by sailors because of a perennial, slowly rotating high-pressure vortex of hot equatorial air that inhales wind and never gives it back. Beneath it, the water describes lazy, clockwise whorls toward a depression at the center.

Its correct name is the North Pacific Subtropical Gyre, though Moore soon learned that oceanographers had another label for it: the Great Pacific Garbage Patch. Captain Moore had wandered into a sump where nearly everything that blows into the water from half the Pacific Rim eventually ends up, spiraling slowly toward a widening horror of industrial excretion. For a week, Moore and his crew found themselves crossing a sea the size of a small continent, covered with floating refuse. It was not unlike an Arctic vessel pushing through chunks of brash ice, except what was bobbing around them was a fright of cups, bottle caps, tangles of fish netting and monofilament line, bits of polystyrene packaging, six-pack rings, spent balloons, filmy scraps of sandwich wrap, and limp plastic bags that defied counting.

Just two years earlier, Moore had retired from his wood-furniture-finishing business. A lifelong surfer, his hair still ungrayed, he'd built himself a boat and settled into what he planned to be a stimulating young retirement. Raised by a sailing father and certified as a captain by the U.S. Coast Guard, he started a volunteer marine environmental monitoring group. After his hellish mid-Pacific encounter with the Great Pacific Garbage Patch, his group ballooned into what is now the Algita Marine Research Foundation, devoted to confronting the flotsam of a half century, since 90 percent of the junk he was seeing was plastic.

What stunned Charles Moore most was learning where it came from. In 1975, the U.S. National Academy of Sciences had estimated that all oceangoing vessels together dumped 8 million pounds of plastic annually. More recent research showed the world's merchant fleet alone shamelessly tossing around 639,000 plastic containers every day. But littering by all the commercial ships and navies, Moore discovered, amounted to mere

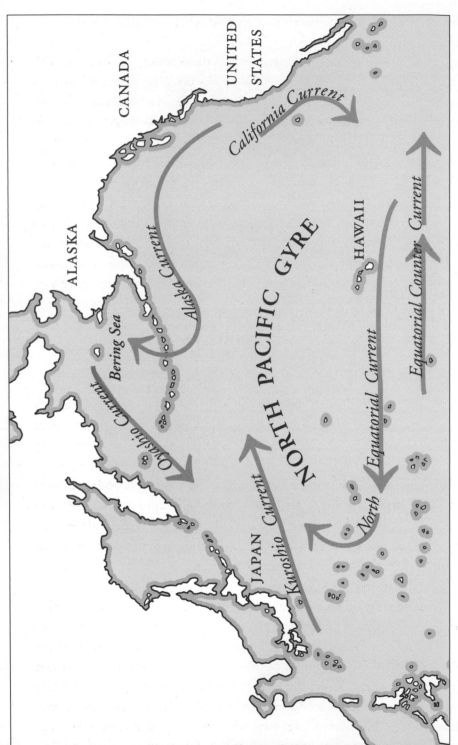

Map of North Pacific Gyre.

MAP BY VIRGINIA NOREY

polymer crumbs in the ocean compared to what was pouring from the shore.

The real reason that the world's landfills weren't overflowing with plastic, he found, was because most of it ends up in an ocean-fill. After a few years of sampling the North Pacific gyre, Moore concluded that 80 percent of mid-ocean flotsam had originally been discarded on land. It had blown off garbage trucks or out of landfills, spilled from railroad shipping containers and washed down storm drains, sailed down rivers or wafted on the wind, and found its way to this widening gyre.

"This," Captain Moore tells his passengers, "is where all the things end up that flow down rivers to the sea." It is the same phrase geologists have uttered to students since the beginning of science, describing the inexorable processes of erosion that reduce mountains to dissolved salts and specks small enough to wash to the ocean, where they settle into layers of the distant future's rocks. However, what Moore refers to is a type of runoff and sedimentation that the Earth had hitherto never known in 5 billion years of geologic time—but likely will henceforth.

During his first 1,000-mile crossing of the gyre, Moore calculated half a pound for every 100 square meters of debris on the surface, and arrived at 3 million tons of plastic. His estimate, it turned out, was corroborated by U.S. Navy calculations. It was the first of many staggering figures he would encounter. And it only represented *visible* plastic: an indeterminate amount of larger fragments get fouled by enough algae and barnacles to sink. In 1998, Moore returned with a trawling device, such as Sir Alistair Hardy had employed to sample krill, and found, incredibly, more plastic by weight than plankton on the ocean's surface.

In fact, it wasn't even close: six times as much.

When he sampled near the mouths of Los Angeles creeks that emptied into the Pacific, the numbers rose by a factor of 100, and kept rising every year. By now he was comparing data with University of Plymouth marine biologist Richard Thompson. Like Thompson, what especially shocked him were plastic bags and the ubiquitous little raw plastic pellets. In India alone, 5,000 processing plants were producing plastic bags. Kenya was churning out 4,000 tons of bags a month, with no potential for recycling.

As for the little pellets known as nurdles, 5.5 *quadrillion*—about 250 billion pounds—were manufactured annually. Not only was Moore find-ing them everywhere, but he was unmistakably seeing the plastic resin bits trapped inside the transparent bodies of jellyfish and salps, the ocean's most prolific and widely distributed filter-feeders. Like seabirds, they'd mistaken brightly colored pellets for fish eggs, and tan ones for krill. And now God-knows-how-many quadrillion little pieces more, coated in body-scrub chemicals and perfectly bite-sized for the little creatures that bigger creatures eat, were being flushed seaward.

What did this mean for the ocean, the ecosystem, the future? All this plastic had appeared in barely more than 50 years. Would its chemical con-stituents or additives—for instance, colorants such as metallic copper—concentrate as they ascended the food chain, and alter evolution? Would it last long enough to enter the fossil record? Would geologists millions of years hence find Barbie doll parts embedded in conglomerates formed in seabed depositions? Would they be intact enough to be pieced together like dinosaur bones? Or would they decompose first, expelling hydrocar-bons that would seep out of a vast plastic Neptune's graveyard for eons to come, leaving fossilized imprints of Barbie and Ken hardened in stone for eons beyond?

Moore and Thompson began consulting materials experts. Tokyo University geochemist Hideshige Takada, who specialized in EDCs—endocrine-disrupting chemicals, or "gender benders"—had been on a gruesome mission to personally research exactly what evils were leaching from garbage dumps all around Southeast Asia. Now he was examining plastic pulled from the Sea of Japan and Tokyo Bay. He reported that in the sea, nurdles and other plastic fragments acted both as magnets and as sponges for resilient poisons like DDT and PCBs.

The use of aggressively toxic polychlorinated biphenyls—PCBs—to make plastics more pliable had been banned since 1970; among other hazards, PCBs were known to promote hormonal havoc such as her-maphroditic fish and polar bears. Like time-release capsules, pre-1970 plastic flotsam will gradually leak PCBs into the ocean for centuries. But, as Takada also discovered, free-floating toxins from all kinds of sources—copy paper, automobile grease, coolant fluids, old fluorescent tubes, and infamous discharges by General Electric and Monsanto plants directly into streams and rivers—readily stick to the surfaces of free-floating plastic.

One study directly correlated ingested plastics with PCBs in the fat tissue of puffins. The astonishing part was the amount. Takada and his colleagues found that plastic pellets that the birds ate concentrate poisons to levels as high as 1 million times their normal occurrence in seawater.

By 2005, Moore was referring to the gyrating Pacific dump as 10 million square miles—nearly the size of Africa. It wasn't the only one: the planet has six other major tropical oceanic gyres, all of them swirling with ugly debris. It was as if plastic exploded upon the world from a tiny seed after World War II and, like the Big Bang, was still expanding. Even if all production suddenly ceased, an astounding amount of the astoundingly durable stuff was already out there. Plastic debris, Moore believed, was now the most common surface feature of the world's oceans. How long would it last? Were there any benign, less-immortal substitutes that civilization could convert to, lest the world be plastic-wrapped evermore?

That fall, Moore, Thompson, and Takada convened at a marine plastic summit in Los Angeles with Dr. Anthony Andrady. A senior research scientist at North Carolina's Research Triangle, Andrady is from Sri Lanka, one of South Asia's rubber-producing powers. While studying polymer science in graduate school, he was distracted from a career in rubber by the surging plastics industry. An 800-page tome he eventually compiled, *Plastics in the Environment,* won him acclaim from the industry and environmentalists alike as the oracle on its subject.

The long-term prognosis for plastic, Andrady told assembled marine scientists, is exactly that: long-term. It's no surprise that plastics have made an enduring mess in the oceans, he explained. Their elasticity, versatility (they can either sink or float), near invisibility in water, durability, and superior strength were exactly why net and fishing line manufacturers had abandoned natural fibers for synthetics such as nylon and polyethylene. In time, the former disintegrate; the latter, even when torn and lost, continue "ghost fishing." As a result, virtually every marine species, including whales, is in danger of being snared by great tangles of nylon loose in the oceans.

Like any hydrocarbon, Andrady said, even plastics "inevitably must biodegrade, but at such a slow rate that it is of little practical consequence. They can, however, photodegrade in a meaningful time frame."

He explained: When hydrocarbons biodegrade, their polymer molecules are disassembled into the parts that originally combined to create them: carbon dioxide and water. When they *photodegrade,* ultraviolet solar radiation weakens plastic's tensile strength by breaking its long, chain-like polymer molecules into shorter segments. Since the strength of plastics depends on the length of their intertwined polymer chains, as the UV rays snap them, the plastic starts to decompose.

Everyone has seen polyethylene and other plastics turn yellow and brittle and start to flake in sunlight. Often, plastics are treated with additives to make them more UV-resistant; other additives can make them more UV-sensitive. Using the latter for six-pack rings, Andrady suggested, might save the lives of many sea creatures.

However, there are two problems. For one, plastic takes much longer to photodegrade in water. On land, plastic left in the sun absorbs infrared heat, and is soon much hotter than the surrounding air. In the ocean, not only does it stay cooled by water, but fouling algae shield it from sunlight.

The other hitch is that even though a ghost fishnet made from photodegradable plastic might disintegrate before it drowns any dolphins, its chemical nature will not change for hundreds, perhaps thousands of years.

"Plastic is still plastic. The material still remains a polymer. Polyethylene is not biodegraded in any practical time scale. There is no mechanism in the marine environment to biodegrade that long a molecule." Even if photodegradable nets helped marine mammals live, he concluded, their powdery residue remains in the sea, where the filter feeders will find it.

"Except for a small amount that's been incinerated," says Tony Andrady the oracle, "every bit of plastic manufactured in the world for the last 50 years or so still remains. It's somewhere in the environment."

That half-century's total production now surpasses 1 billion tons. It includes hundreds of different plastics, with untold permutations involving added plasticizers, opacifiers, colors, fillers, strengtheners, and light stabilizers. The longevity of each can vary enormously. Thus far, none has disappeared. Researchers have attempted to find out how long it will take polyethylene to biodegrade by incubating a sample in a live bacteria culture. A year later, less than 1 percent was gone.

"And that's under the best controlled laboratory conditions. That's not what you will find in real life," says Tony Andrady. "Plastics haven't been around long enough for microbes to develop the enzymes to handle it, so they can only biodegrade the very-low-molecular-weight part of the plastic"—meaning, the smallest, already-broken polymer chains. Although truly biodegradable plastics derived from natural plant sugars have appeared, as well as biodegradable polyester made from bacteria, the chances of them replacing the petroleum-based originals aren't great.

"Since the idea of packaging is to protect food from bacteria," Andrady observes, "wrapping leftovers in plastic that encourages microbes to eat it may not be the smartest thing to do."

But even if it worked, or even if humans were gone and never produced another nurdle, all the plastic already produced would remain—how long?

"Egyptian pyramids have preserved corn, seeds, and even human parts such as hair because they were sealed away from sunlight with little oxygen or moisture," says Andrady, a mild, precise man with a broad face and a clipped, persuasively reasonable voice. "Our waste dumps are somewhat like that. Plastic buried where there's little water, sun, or oxygen will stay intact a long time. That is also true if it is sunk in the ocean, covered with sediment. At the bottom of the sea, there's no oxygen, and it's very cold."

He gives a clipped little laugh. "Of course," he adds, "we don't know much about microbiology at those depths. Possibly anaerobic organisms there can biodegrade it. It's not inconceivable. But no one's taken a submersible down to check. Based on our observations, it's unlikely. So we expect much-slower degradation at the sea bottom. Many times longer. Even an order of magnitude longer."

An order of magnitude—that's 10 times—longer than what? One thousand years? Ten thousand?

No one knows, because no plastic has died a natural death yet. It took today's microbes that break hydrocarbons down to their building blocks a long time after plants appeared to learn to eat lignin and cellulose. More recently, they've even learned to eat oil. None can digest plastic yet, because 50 years is too short a time for evolution to develop the necessary biochemistry.

"But give it 100,000 years," says Andrady the optimist. He was in his native Sri Lanka when the Christmas 2004 tsunami hit, and even there, after those apocalyptic waters struck, people found reason to hope. "I'm sure you'll find many species of microbes whose genes will let them do this tremendously advantageous thing, so that their numbers will grow and prosper. Today's amount of plastic will take hundreds of thousands of years to consume, but, eventually, it will all biodegrade. Lignin is far more complex, and it biodegrades. It's just a matter of waiting for evolution to catch up with the materials we are making."

And should biologic time run out and some plastics remain, there is always geologic time.

"The upheavals and pressure will change it into something else. Just like trees buried in bogs a long time ago—the geologic process, not biodegradation, changed them into oil and coal. Maybe high concentrations of plastics will turn into something like that. Eventually, they will change. Change is the hallmark of nature. Nothing remains the same."

CHAPTER 10

❧

The Petro Patch

WHEN HUMANS DEPART, among the immediate beneficiaries of our absence will be mosquitoes. Although our anthropocentric worldview may flatter us into thinking that human blood is essential to their survival, in fact they are versatile gourmets capable of supping at the veins of most warm-blooded mammals, cold-blooded reptiles, and even birds. In our absence, presumably plenty of wild and feral creatures will rush to fill our void and set up house in our abandoned spaces. Their numbers no longer culled by our lethal traffic, they should multiply with such abandon that humanity's total biomass—which the eminent biologist E. O. Wilson estimates wouldn't fill the Grand Canyon—won't be missed for long.

At the same time, any mosquitoes still bereaved by our passing will be consoled by two bequests. First, we'll stop exterminating them. Humans were targeting mosquitoes long before the invention of pesticides, by spreading oil on the surfaces of ponds, estuaries, and puddles where they breed. This larvicide, which denies baby mosquitoes oxygen, is still widely practiced, as are all other manners of antimosquito chemical warfare. They range from hormones that keep larvae from maturing into adults, to—especially in the malarial tropics—aerial spraying of DDT, banned only in parts of the world. With humans gone, billions of the little buzzers that would otherwise have died prematurely will now live, and among the secondary beneficiaries will be many freshwater fish species, in whose food chains mosquito eggs and larvae form big links. Others will be flowers:

when mosquitoes aren't sucking blood, they sip nectar—the main meal for all male mosquitoes, although vampirish females drink it as well. That makes them pollinators, so the world without us will bloom anew.

The other gift to mosquitoes will be restoration of their traditional homelands—in this case, home waters. In the United States alone, since its founding in 1776, the part of their prime breeding habitat, wetlands, that they have lost equals twice the area of California. Put that much land back into swamps, and you get the idea. (Mosquito population growth would have to be adjusted for corresponding increases in mosquito-eating fish, toads, and frogs—though, with the last two, humans may have given the insects yet another break: It's unclear how many amphibians will survive chytrid, an escaped fungus spread by the international trade in laboratory frogs. Triggered by rising temperatures, it has annihilated hundreds of species worldwide.)

Habitat or not, as anyone knows who lives atop a former marsh that was drained and developed, be it in suburban Connecticut or a Nairobi slum, mosquitoes always find a way. Even a dew-filled plastic bottle cap can incubate a few of their eggs. Until asphalt and pavement decompose for good and wetlands rise up to reclaim their former surface rights, mosquitoes will make do with puddles and backed-up sewers. And they can also rest assured that one of their favorite man-made nurseries will be intact for, at minimum, another century, and will continue making cameo appearances for many more centuries thereafter: scrapped rubber automobile tires.

Rubber is a kind of polymer called an elastomer. The ones that occur in nature, such as the milky latex extract of the Amazonian Pará tree, are, logically, biodegradable. The tendency of natural latex to turn gooey in high temperatures, and to stiffen or even shatter in cold, limited its practicality until 1839, when a Massachusetts hardware salesman tried mixing it with sulfur. When he accidentally dropped some on a stove and it didn't melt, Charles Goodyear realized that he'd created something that nature had never tried before.

To this day, nature hasn't come up with a microbe that eats it, either. Goodyear's process, called vulcanization, ties long rubber polymer chains together with short strands of sulfur atoms, actually transforming them into a single giant molecule. Once rubber is vulcanized—meaning it's heated, spiked with sulfur, and poured into a mold, such as one shaped like a truck tire—the resulting huge molecule takes that form and never relinquishes it.

Being a single molecule, a tire can't be melted down and turned into something else. Unless physically shredded or worn down by 60,000 miles of friction, both entailing significant energy, it remains round. Tires drive landfill operators crazy, because when buried, they encircle a doughnut-shaped air bubble that wants to rise. Most garbage dumps no longer accept them, but for hundreds of years into the future, old tires will inexorably work their way to the surface of forgotten landfills, fill with rainwater, and begin breeding mosquitoes again.

In the United States, an average of one tire per citizen is discarded annually—that's a third of a billion, just in one year. Then there's the rest of the world. With about 700 million cars currently operating and far more than that already junked, the number of used tires we'll leave behind will be less than a trillion, but certainly many, many billions. How long they'll lie around depends on how much direct sunlight falls on them. Until a microbe evolves that likes its hydrocarbons seasoned with sulfur, only the caustic oxidation of ground-level ozone, the pollutant that stings your sinuses, or the cosmic power of ultraviolet rays that penetrate a damaged stratospheric ozone layer, can break vulcanized sulfur bonds. Automobile tires therefore are impregnated with UV inhibitors and "anti-ozonants," along with other additives like the carbon black filler that gives tires their strength and color.

With all that carbon in tires, they can be also burned, releasing considerable energy, which makes them hard to extinguish, along with surprising amounts of oily soot that contains some noxious components we invented in a hurry during World War II. After Japan invaded Southeast Asia, it controlled nearly the entire world's rubber supply. Understanding that their own war machines wouldn't go far using leather gaskets or wooden wheels, both Germany and the United States drafted their top industries to find a substitute.

The largest plant in the world today that produces synthetic rubber is in Texas. It belongs to the Goodyear Tire & Rubber Company and was built in 1942, not long after scientists figured out how to make it. Instead of living tropical trees, they used dead marine plants: phytoplankton that died 300 million to 350 million years ago and sank to the sea bottom.

Eventually—so the theory goes; the process is poorly understood and sometimes challenged—the phytoplankton were covered with so many sediments and squeezed so hard they metamorphosed into a viscous liquid. From that crude oil, scientists already knew how to refine several useful hydrocarbons. Two of these—styrene, the stuff of Styrofoam, and butadiene, an explosive and highly carcinogenic liquid hydrocarbon—provided the combination to synthesize rubber.

Six decades later, it's what Goodyear Rubber still makes here, with the same equipment rolling out the base for everything from NASCAR racing tires to chewing gum. Large as the plant is, however, it's swamped by what surrounds it: one of the most monumental constructs that human beings have imposed on the planet's surface. The industrial megaplex that begins on the east side of Houston and continues uninterrupted to the Gulf of Mexico, 50 miles away, it is the largest concentration of petroleum refineries, petrochemical companies, and storage structures on Earth.

It contains, for example, the tank farm behind razor-edged concertina wire just across the highway from Goodyear—a cluster of cylindrical crude-oil receptacles each a football field's length in diameter, so wide they appear squat. The omnipresent pipelines that link them run to all compass points, as well as up and down—white, blue, yellow, and green pipelines, the big ones nearly four feet across. At plants like Goodyear, pipelines form archways high enough for trucks to pass under.

Those are just the visible pipes. A satellite-mounted CT scanner flying over Houston would reveal a vast, tangled, carbon-steel circulatory system about three feet below the surface. As in every city and town in the developed world, thin capillaries run down the center of every street, branching off to every house. These are gas lines, comprising so much steel that it's a wonder that compass needles don't simply point toward the ground. In Houston, however, gas lines are mere accents, little flourishes. Refinery pipelines wrap around the city as tightly as a woven basket. They move material called light fractions, distilled or catalytically cracked off crude oil, to hundreds of Houston chemical plants—such as Texas Petrochemical, which provides its neighbor Goodyear with butadiene and also concocts a related substance that makes plastic wrap cling. It also produces butane—the feedstock for polyethylene and polypropylene nurdle pellets.

Hundreds of other pipes full of freshly refined gasoline, home heating oil, diesel, and jet fuel hook into the grand patriarch of conduits—the

5,519-mile, 30-inch Colonial Pipeline, whose main trunk starts in the Houston suburb of Pasadena. It picks up more product in Louisiana, Mississippi, and Alabama, then climbs the eastern seaboard, sometimes above-ground, sometimes below. The Colonial is typically filled with a lineup of various grades of fuel that pump through it at about four miles per hour until they're disgorged at a Linden, New Jersey, terminal just below New York Harbor—about a 20-day trip, barring shutdowns or hurricanes.

Imagine future archaeologists clanging their way through all those pipes. What will they make of the thick old steel boilers and multiple stacks behind Texas Petrochemical? (Although, if humans stick around for a few more years, all that old stock, overbuilt back when there were no computers to pinpoint tolerances, will have been dismantled and sold to China, which is buying up scrap iron in America for purposes that some World War II historians question with alarm.)

If those archaeologists were to follow the pipes several hundred feet down, they would encounter an artifact destined to be among the longest-lasting ever made by humans. Beneath the Texas Gulf coast are about 500 salt domes formed when buoyant salts from saline beds five miles down rise through sedimentary layers. Several lie right under Houston. Bullet-shaped, they can be more than a mile across. By drilling into a salt dome and then pumping in water, it is possible to dissolve its interior and use it for storage.

Some salt dome storage caverns below the city are 600 feet across and more than half a mile tall, equaling a volume twice that of the Houston Astrodome. Because salt crystal walls are considered impermeable, they are used for storing gases, including some of the most explosive, such as eth-ylene. Piped directly to an underground salt dome formation, ethylene is stored under 1,500 pounds of pressure until it's ready to be turned into plastic. Because it is so volatile, ethylene can decompose rapidly and blow a pipe right out of the ground. Presumably, it would be best for archaeol-ogists of the future to leave the salt caverns be, lest an ancient relic from a long-dead civilization blow up in their faces. But how would they know?

Back above ground, like robotic versions of the mosques and minarets that grace the shores of Istanbul's Bosphorus, Houston's petroscape of domed white tanks and silver fractionating towers spreads along the banks of its

Ship Channel. The flat tanks that store liquid fuels at atmospheric temperatures are grounded so that vapors that gather in the space below the roof don't ignite during a lightning storm. In a world without humans to inspect and paint doubled-hulled tanks, and replace them after their 20-year life span, it would be a race to see whether their bottoms corrode first, spilling their contents into the soil, or their grounding connectors flake away—in which case, explosions would hasten deterioration of the remaining metal fragments.

Some tanks with moveable roofs that float atop liquid contents to avoid vapor buildup might fail even earlier, as their flexible seals start to leak. If so, what's inside would just evaporate, pumping the last remaining human-extracted carbon into the atmosphere. Compressed gases, and some highly inflammable chemicals such as phenols, are held in spherical tanks, which should last longer because their hulls aren't in contact with the ground—although, since they're pressurized, they would explode more sensationally once their spark protection rusts away.

What lies beneath all this hardware, and what are the chances that it could ever recover from the metallic and chemical shock that the last century of petrochemical development has wreaked here? Should this most unnatural of all Earthly landscapes ever be abandoned by the humans who keep its flares burning and fuels flowing, how could nature possibly dismantle, let alone decontaminate, the great Texas petroleum patch?

HOUSTON, ALL 620 square miles of it, straddles the edge between a bluestem and grama-grass prairie that once grew belly-high to a horse and the lower piney-woods wetland that was (and still is) part of the original delta of the Brazos River. The dirt-red Brazos begins far across the state, draining New Mexico mountains 1,000 miles away, then cuts through Texas hill country and eventually dumps one of the biggest silt loads on the continent into the Gulf of Mexico. During glacial times, when winds blowing off the ice sheet slammed into warm gulf air and caused torrential rains, the Brazos laid down so much sediment that it would dam itself and as a result slip back and forth across a deltaic fan hundreds of miles wide. Lately, it passes just south of town. Houston sits along one of the river's former channels, atop 40,000 feet of sedimentary clay deposits.

In the 1830s, that magnolia-lined channel, Buffalo Bayou, attracted entrepreneurs who noticed that it was navigable from Galveston Bay to the edge of the prairie. At first, the new town they built there shipped cotton 50 miles down this inland waterway to the port of Galveston, then the biggest city in Texas. After 1900, when the deadliest hurricane in U.S. history hit Galveston and killed 8,000 people, Buffalo Bayou was widened and deepened into the Ship Channel, to make Houston a seaport. Today, by cargo volume, it's America's biggest, and Houston itself is huge enough to hold Cleveland, Baltimore, Boston, Pittsburgh, Denver, and Washington, D.C., with room to spare.

Galveston's misfortune coincided with discoveries of oil along the Texas Gulf coast and the advent of the automobile. Longleaf piney woods, bottomland delta hardwood forests, and coastal prairie soon were supplanted by drilling rigs and dozens of refineries along Houston's shipping corridor. Next came chemical plants, then World War II rubber factories, and, finally, the fabulous postwar plastics industry. Even when Texas oil production peaked in the 1970s and then plummeted, Houston's infrastructure was so vast that the world's crude kept flowing here to be refined.

The tankers, bearing flags of Middle Eastern nations, Mexico, and Venezuela, arrive at an appendage of the Ship Channel on Galveston Bay called Texas City, a town of about 50,000 that has as much acreage devoted to refining as to residences and business. Compared to their big neighbors—Sterling Chemical, Marathon, Valero, BP, ISP, Dow—the bungalows of Texas City's residents, mostly black and Latino, are lost in a townscape ruled by the geometry of petrochemistry: circles, spheres, and cylinders—some tall and thin, some short and flat, some wide and round.

It is the tall ones that tend to blow up.

Not all of them, although they often look alike. Some are wet-gas scrubbers: towers that use Brazos River water to quench gas emissions and cool down hot solids, generating white steam clouds up their stacks. Others are fractionating towers, in which crude oil is heated from the bottom to distill it. The various hydrocarbons in crude, ranging from tar to gasoline to natural gas, have different boiling points; as they're heated, they separate, arranging themselves in the column with the lightest ones on top.

As long as expanding gases are drawn off to release pressure, or the heat is eventually reduced, it's a fairly safe process.

Trickier are the ones that add other chemicals to convert petroleum into something new. In refineries, catalytic cracking towers heat the heavy hydrocarbons with a powdered aluminum silicate catalyst to about 1,200°F. This literally cracks their big polymer chains into smaller, lighter ones, such as propane or gasoline. Injecting hydrogen into the process can produce jet fuel and diesel. All these, especially at high temperatures, and especially with hydrogen involved, are highly explosive.

A related procedure, isomerization, uses a platinum catalyst and even more heat to rearrange atoms in hydrocarbon molecules for boosting fuel octane or making substances used in plastics. Isomerization can get extremely volatile. Connected to these cracking towers and isomerization plants are flares. If any process becomes imbalanced or if temperatures shoot too high, flares are there to bleed off pressure. A release valve sends whatever can't be contained up the flare stack, signaling a pilot to ignite. Sometimes steam is injected so that whatever it is doesn't smoke, but burns cleanly.

When something malfunctions, the results, unfortunately, can be spectacular. In 1998, Sterling Chemical expelled a cloud of various benzene isomers and hydrochloric acid that hospitalized hundreds. That followed a leak of 3,000 pounds of ammonia four years earlier that prompted 9,000 personal injury suits. In March 2005, a geyser of liquid hydrocarbons erupted from one of BP's isomerization stacks. When it hit the air, it ignited and killed 15 people. That July, at the same plant, a hydrogen pipe exploded; in August, a gas leak reeking of rotten eggs, which signals toxic hydrogen sulfide, shut much of BP down for a while. Days later, at a BP plastics-manufacturing subsidiary 15 miles south at Chocolate Bayou, flames exploded 50 feet in the air. The blaze had to be left to burn itself out. It took three days.

The oldest refinery in Texas City, started in 1908 by a Virginia farmers' cooperative to produce fuel for their tractors, is owned today by Valero Energy Corporation. In its modern incarnation, it has earned one of the highest safety designations among U.S. refineries, but it is still a place designed to draw energy from a crude natural resource by transmuting it into more explosive forms. That energy feels barely contained by Valero's humming labyrinth of valves, gauges, heat exchangers, pumps, absorbers, separators, furnaces, incinerators, flanges, tanks girdled by spiral stairwells, and serpentine loops of red, yellow, green, and silvery pipes (the silver ones are insulation-wrapped, meaning that something inside is hot, and needs to stay that way). Looming overhead are 20 fractionation towers and 20 more exhaust stacks. A coker shovel, basically a crane with a bucket on it, shuttles back and forth, dumping loads of sludge redolent of asphalt—the heavy ends of the crude spectrum, left in the bottom of the fractionators—onto conveyors leading into a catalytic cracker, to squeeze another barrel of diesel from them.

Above all this are the flares, wedges of flame against a whitish sky, keeping all the organic chemistry in equilibrium by burning off pressures that build faster than all the monitoring gauges can regulate them. There are gauges that read the thickness of steel pipe at the right-angle bends where hot, corrosive fluids smash, to predict when they will fail. Anything that contains hot liquid traveling at high speeds can develop stress cracks, especially when the liquid is heavy crude, laden with metals and sulfur that can eat pipe walls.

All this equipment is controlled by computers—until something exceeds what the computer can correct. Then the flares kick in. Suppose, though, that a system's pressures exceeded their capacity—or suppose nobody were around to notice the overload. Normally, somebody always is, around the clock. But what if human beings suddenly disappeared while the plant was still operating?

"You'd end up with a break in some vessel," says Valero spokesman Fred Newhouse, a compact, congenial man with light brown skin and grizzled hair. "And probably a fire." But at that point, Newhouse adds, fail-safe control valves up and downstream from the accident would automatically trip. "We measure pressure, flow, and temperature constantly. Any changes would isolate the problem so that fire couldn't ripple from that unit to the next one."

But what if no one were left to fight the flames? And what if all the power died, because no one was manning any of the coal, gas, and nuclear plants, or any of the hydroelectric dams from California to Tennessee, all of which funnel electrons through a Houston grid connection to keep the lights on in Texas City? And what if the automatic emergency generators ran out of diesel, so no signal tripped the shutoff valves?

Newhouse moves into the shadow of a cracking tower to consider this. After 26 years at Exxon, he really likes working for Valero. He's proud of their clean record, especially compared to the BP plant across the road, which the EPA in 2006 named the nation's worst polluter. The thought of all this incredible infrastructure out of control, torching itself, makes him wince.

"Okay. Everything would burn until all the hydrocarbon in the system was gone. But," he insists, "it's very unlikely that fire would spread beyond the property. The pipes that connect the Texas City refineries all have check valves to isolate one from another. So even when you see plants explode," he says, gesturing across the road, "adjacent units aren't damaged. Even if it's a huge fire, fail-safe systems are in place."

E.C. isn't so sure. "Even on a normal operating day," he says, "a petrochemical plant is a ticking time bomb." A chemical plant and refinery inspector, he's seen volatile light petroleum fractions do some interesting things on their way to becoming secondary petrochemicals. When light-end chemicals such as ethylene or acrylonitrile—a highly inflammable precursor of acrylic, hazardous to human nervous systems—are under high pressure, they often slip through ducts and find their way to adjacent units, or even adjacent refineries.

In the event that humans were gone tomorrow, he says, what would happen to petroleum refineries and chemical plants would depend on whether anybody bothered to flip some switches before departing.

"Supposing there's time for a normal shutdown. High pressures would be brought down to low pressure. Boilers would be shut down, so temperature isn't a problem. In the towers, the heavy bottoms would cake up into solid goop. They would be encased in vessels with steel inner layers, surrounded by Styrofoam or glass-fiber insulation, with an outer skin of sheet metal. Between those layers there's often steel or copper tubing filled with water to control temperatures. So whatever is in them would be stable— until corrosion set in from the soft water."

He rummages in a desk drawer, then closes it. "Absent any fire or explosion, light-end gases will dissipate into the air. Any sulfur by-product lying around will eventually dissolve and create acid rain. Ever see a Mexican refinery? There're mountains of sulfur. Americans ship it off. Anyhow, refineries also have big tanks of hydrogen. Very volatile, but if they leaked, the hydrogen would float away. Unless lightning blew it up first."

He laces his fingers behind his curly, graying brown hair and tilts back in his office chair. "Now that would get rid of lot of cement infrastructure right there."

And if there were no time to shut down a plant, if humans were raptured off to heaven or another galaxy and left everything running?

He rocks forward. "At first, emergency power plants would kick in. They're usually diesel. They would probably maintain stability until they depleted their fuel. Then you'd have high pressures and high temperatures. With no one to monitor controls or the computers, some reactions would run away and go boom. You would get a fire, and then a domino effect, since there'd be nothing to stop it. Even with emergency motors, water sprayers wouldn't work, because there'd be no one to turn them on. Some relief valves would vent, but in a fire, a relief valve would just feed the flames."

E.C. swivels completely in his chair. A marathoner, he wears jogging shorts and a sleeveless T-shirt. "All the pipes would be conduits for fires. You'd have gas going from one area to another. Normally, in emergencies you shut down the connections, but none of that would happen. Things would just spread from one facility to the next. That blaze could possibly go for weeks, ejecting stuff into the atmosphere."

Another swivel, this time counterclockwise. "If this happened to every plant in world, imagine the amount of pollutants. Think of the Iraqi fires. Then multiply that, everywhere."

In those Iraqi fires, Saddam Hussein blew up hundreds of wellheads, but sabotage isn't always needed. Mere static electricity from fluids moving through pipes can spark ignition in natural-gas wells, or in oil wells pressurized with nitrogen to bubble up more petroleum. On the big flat screen in front of E.C., a blinking item on a list says that a Chocolate Bayou, Texas, plant that makes acrylonitrile was 2002's biggest releaser of carcinogens in the United States.

"Look: If all the people left, a fire in a gas well would go until the gas pocket depleted. Usually, the ignition sources are wiring, or a pump. They'd be dead, but you'd still have static electricity or lightning. A well fire burns on the surface, since it needs air, but there would be no one to push it back and cap the wellhead. Huge pockets of gas in the Gulf of

Mexico or Kuwait would maybe burn forever. A petrochemical plant wouldn't go that long, because there's not as much to burn. But imagine a runaway reaction with burning plants throwing up clouds of stuff like hydrogen cyanide. There would be a massive poisoning of the air in the Texas-Louisiana chemical alley. Follow the trade winds and see what happens."

All those particulates in the atmosphere, he imagines, could create a mini chemical nuclear winter. "They would also release chlorinated compounds like dioxins and furans from burning plastics. And you'd get lead, chromium, and mercury attached to the soot. Europe and North America, with the biggest concentrations of refineries and chemical plants, would be the most contaminated. But the clouds would disperse through the world. The next generation of plants and animals, the ones that didn't die, might need to mutate in ways that could impact evolution."

ON THE NORTHERN edge of Texas City, in the long afternoon shadow of an ISP chemical plant, is a 2,000-acre wedge of original tall grass donated by Exxon-Mobil and now managed by The Nature Conservancy. It is the last remnant of what were 6 million acres of coastal prairie before petroleum arrived. Today, the Texas City Prairie Preserve is home to half of the 40 known remaining Attwater's prairie chickens—considered the most endangered bird in North America until the controversial 2005 spotting in Arkansas of a lone ivory-billed woodpecker, a species hitherto believed extinct.

During courtship, male Attwater's prairie chickens inflate vivid, balloon-like golden sacs on either side of their necks. The impressed females respond by laying a lot of eggs. In a world without humans, however, it's questionable whether the breed will be able to survive. Oil industry apparatus isn't all that has spread across their habitat. The grassland here once ran clear to Louisiana with hardly a tree, the tallest thing on the horizon being an occasional grazing buffalo. That changed around 1900 with the coincidental arrival of both petroleum and the Chinese tallow tree.

Back in China, this formerly cold-weather species coated its seeds with harvestable quantities of wax to guard against winter. Once it was brought to the balmy American South as an agricultural crop, it noticed there was no need to do that. In a textbook display of sudden evolutionary adaptation, it stopped making weatherproof wax and put its energy into producing more seeds.

Today, wherever there isn't a petrochemical stack along the Ship Channel, there's a Chinese tallow tree. Houston's longleaf pines are long gone, overwhelmed by the Chinese interloper, its rhomboid leaves turning ruby red each fall in atavistic memory of chilly Canton. The only way The Nature Conservancy keeps them from shading out and shoving aside the bluestem and sunflowers of its prairie is with careful annual burning to keep the prairie chicken mating fields intact. Without people to maintain that artificial wilderness, only an occasional exploding old petroleum tank might beat back the botanical Asian invasion.

If, in the immediate aftermath of *Homo sapiens petrolerus,* the tanks and towers of the Texas petrochemical patch all detonated together in one spectacular roar, after the oily smoke cleared, there would remain melted roads, twisted pipe, crumpled sheathing, and crumbled concrete. White-hot incandescence would have jump-started the corrosion of scrap metals in the salt air, and the polymer chains in hydrocarbon residues would likewise have cracked into smaller, more digestible lengths, hastening biodegradation. Despite the expelled toxins, the soils would also be enriched with burnt carbon, and after a year of rains switchgrass would be growing. A few hardy wildflowers would appear. Gradually, life would resume.

Or, if the faith of Valero Energy's Fred Newhouse in system safeguards proves warranted—or if the departing oilmen's last loyal act is to depressurize towers and bank the fires—the disappearance of Texas's world champion petroleum infrastructure will proceed more slowly. During the first few years, the paint that slows corrosion will go. Over the next two decades, all the storage tanks will exceed their life spans. Soil moisture, rain, salt, and Texas wind will loosen their grip until they leak. Any heavy crude will have hardened by then; weather will crack it, and bugs will eventually eat it.

What liquid fuels that haven't already evaporated will soak into the ground. When they hit the water table, they'll float on top because oil is lighter than water. Microbes will find them, realize that they were once only plant life, too, and gradually adapt to eat them. Armadillos will return to burrow in the cleansed soil, among the rotting remains of buried pipe.

Unattended oil drums, pumps, pipes, towers, valves, and bolts will deteriorate at the weakest points, their joints. "Flanges, rivets," says Fred Newhouse. "There are a jillion in a refinery." Until they go, collapsing the metal walls, pigeons that already love to nest atop refinery towers will speed the corruption of carbon steel with their guano, and rattlesnakes will nest in the vacant structures below. As beavers dam the streams that trickle into Galveston Bay, some areas will flood. Houston is generally too warm for a freeze-thaw cycle, but its deltaic clay soils undergo formidable swell-shrink bouts as rains come and go. With no more foundation repairmen to shore up the cracks, in less than a century downtown buildings will start leaning.

During that same time, the Ship Channel will have silted back into its former Buffalo Bayou self. Over the next millennium, it and the other old Brazos channels will periodically fill, flood, undermine the shopping malls, car dealerships, and entrance ramps—and, building by tall building, bring down Houston's skyline.

As for the Brazos itself: Today, 20 miles down the coast from Texas City, just below Galveston Island and just past the venomous plumes rising from Chocolate Bayou, the Brazos de Dios ("Arms of God") River wanders around a pair of marshy national wildlife refuges, drops an island's worth of silt, and joins the Gulf of Mexico. For thousands of years, it has shared a delta, and sometimes a mouth, with the Colorado and the San Bernard rivers. Their channels have interbraided so often that the correct answer to which is which is temporary at best.

Much of the surrounding land, barely three feet above sea level, is dense canebrake and old bottomland forest stands of live oaks, ashes, elms, and native pecans, spared years ago by sugarcane plantations for cattle shade. "Old" here means only a century or two, because clay soils repel root penetration, so that mature trees tend to list until the next hurricane knocks them over. Hung with wild grapevines and beards of Spanish moss, these woods are seldom visited by humans, who are dissuaded by poison ivy and black snakes, and also by golden orb weaver spiders big as a human hand, which string viscous webs the size of small trampolines between tree trunks. There are enough mosquitoes to belie any notion that their survival would be threatened when evolving microbes finally bring down the world's mountain ranges of scrap tires.

As a result, these neglected woods are inviting habitats for cuckoos, woodpeckers, and wading birds such as ibises, sandhill cranes, and roseate spoonbills. Cottontail and marsh rabbits attract barn owls and bald eagles, and each spring thousands of returning passerine birds, including scarlet and summer tanagers in fabulous breeding plumage, flop into these trees after a long gulf crossing.

The deep clays below their perches accumulated back when the Brazos flooded—back before a dozen dams and diversions and a pair of canals siphoned its water to Galveston and Texas City. But it will flood again. Untended dams silt up fast. Within a century without humans, the Brazos will spill over all of them, one by one.

It may not even have to wait that long. Not only is the Gulf of Mexico, whose water is even warmer than the ocean's, creeping inland, but all along the Texas coast for the past century, the ground has been lowered to receive it. When oil, gas, or groundwater is pumped from beneath the surface, land settles into the space it occupied. Subsidence has lowered parts of Galveston 10 feet. An upscale subdivision in Baytown, north of Texas City, dropped so low that it drowned during Hurricane Alicia in 1983 and is now a wetlands nature preserve. Little of the Gulf Coast is more than three feet above sea level, and parts of Houston actually dip below it.

Lower the land, raise the seas, add hurricanes far stronger than midsize, Category 3 Alicia, and even before its dams go, the Brazos gets to do again what it did for 80,000 years: like its sister to the east, the Mississippi, it will flood its entire delta, starting up where the prairie ends. Flood the enormous city that oil built, all the way down to the coast. Swallow the San Bernard and overlap the Colorado, fanning a sheet of water across hundreds of miles of coastline. Galveston Island's 17-foot seawall won't be much help. Petroleum tanks along the Ship Channel will be submerged; flare towers, catalytic crackers, and fractionating columns, like downtown Houston buildings, will poke out of brackish floodwaters, their foundations rotting while they wait for the waters to recede.

Having rearranged things yet again, the Brazos will choose a new course to the sea—a shorter one, because the sea will be nearer. New bottomlands will form, higher up, and eventually new hardwoods will appear (assuming that Chinese tallow trees, whose waterproof seeds should make them permanent colonizers, share the riparian space with them). Texas City will be missing; hydrocarbons leaching out of its drowned petro-chemical plants will swirl and dissipate in the currents, with a few heavy-end crude residues dumped as oil globules on the new inland shores, eventually to be eaten.

Below the surface, the oxidizing metal parts of chemical alley will provide a place for Galveston oysters to attach. Silt and oyster shells will slowly bury them, and will then be buried themselves. Within a few million years, enough layers will amass to compress shells into limestone, which will bear an odd, intermittent rusty streak flecked with sparkling traces of nickel, molybdenum, niobium, and chromium. Millions of years after that, someone or something might have the knowledge and tools to recognize the signal of stainless steel. Nothing, however, will remain to suggest that its original form once stood tall over a place called Texas, and breathed fire into the sky.

CHAPTER 11

The World Without Farms

1. The Woods

WHEN WE THINK civilization, we usually picture a city. Small wonder: we've gawked at buildings ever since we started raising towers and temples, like Jericho's. As architecture soared skyward and marched outward, it was unlike anything the planet had ever known. Only beehives or ant mounds, on a far humbler scale, matched our urban density and complexity. Suddenly, we were no longer nomads cobbling ephemeral nests out of sticks and mud, like birds or beavers. We were building homes to last, which meant we were staying in one place. The word *civilization* itself derives from the Latin *civis,* meaning "town dweller."

Yet it was the farm that begat the city. Our transcendental leap to sowing crops and herding critters—actually controlling other living things—was even more world-shaking than our consummate hunting skill. Instead of simply gathering plants or killing animals just prior to eating them, we now choreographed their existence, coaxing them to grow more reliably and far more abundantly.

Since a few farmers could feed many, and since intensified food production meant intensified people production, suddenly there were a lot of humans free to do things other than gather or grow meals. With the possible exception of Cro-Magnon cave artists, who may have been so esteemed for their talents that they were relieved of other duties, until agriculture arrived, food-finding was the only occupation for humans on this planet.

Agriculture let us settle down, and settlement led to urbanity. Yet, imposing as skylines are, farmlands have much more impact. Nearly 12 percent of the planet's landmass is cultivated, compared to about 3 percent occupied by towns and cities. When grazing land is included, the amount of Earthly terrain dedicated to human food production is more than one-third of the world's land surface.

If we suddenly stopped plowing, planting, fertilizing, fumigating, and harvesting; if we ceased fattening goats, sheep, cows, swine, poultry, rabbits, Andean guinea pigs, iguanas, and alligators, would those lands return to their former, pre-agro-pastoral state? Do we even know what that was?

For an idea of how the land on which we've toiled might or might not recover from us, we can begin in two Englands—one old, one New.

In any New England woods south of Maine's boreal wilderness, within five minutes you see it. A forester's or ecologist's trained eye notices it just by spotting a stand of big white pine, which only grow in such uniform density in a former cleared field. Or they spot clusters of hardwoods— beech, maples, oaks—of similar age, which sprouted in the shade of a now-missing stand of white pines that were cut or blown away in a hurricane, leaving hardwood seedlings an open sky to fill with their canopies.

But even if you don't know a birch from a beech, you can't miss seeing it around knee-height, camouflaged by fallen leaves and lichens, or wrapped in green brambles. Someone has been here. The low stone walls that crisscross the forests of Maine, Vermont, New Hampshire, Massachusetts, Connecticut, and upstate New York reveal that humans once staked boundaries here. An 1871 fencing census, writes Connecticut geologist Robert Thorson, showed at least 240,000 miles of handmade stone walls east of the Hudson River—enough to reach to the moon.

As the last glaciers of the Pleistocene advanced, stones were ripped from granite outcroppings, then dropped as they melted back. Some lay on the surface; some were ground into the subsoil, to be periodically heaved up by frost. All had to be cleared along with trees so that transplanted European farmers could start over in a New World. The stones and boulders they moved marked the borders of their fields and penned their animals.

So far from large markets, raising beef wasn't practical, but for their own use New England's farmers kept enough cattle, pigs, and dairy cows that most of their land was pasture and hay fields. The rest was in rye, barley, early wheat, oats, corn, or hops. The trees they downed and the stumps they yanked were of the mixed hardwood, pine, and spruce forests we identify with New England today—and we do, because they're back.

Unlike almost anywhere else on Earth, New England's temperate forest is increasing, and now far exceeds what it was when the United States was founded in 1776. Within 50 years of U.S. independence, the Erie Canal was dug across New York State and the Ohio Territory opened—an area whose shorter winters and loamier soils lured away struggling Yankee farmers. Thousands more didn't bother to return to the soil after the Civil War, but headed instead into factories and mills powered by New England's rivers—or headed west. As the forests of the Midwest began to come down, the forests of New England began coming back.

The unmortared stone walls built by three centuries of farmers flex as soil swells and shrinks with the seasons. They should be part of the landscape for a few more centuries, until the leaf litter turns to more soil and buries them. But how similar are the forests growing around them to what was here before the Europeans arrived, or the Indians before them? And untouched, what would they become?

In his 1980 book *Changes in the Land,* geographer William Cronon challenged historians who wrote of Europeans encountering an unsullied forest primeval when they first arrived in the New World—a forest supposedly so unbroken that a squirrel might leap treetops from Cape Cod to the Mississippi without ever having to touch ground. Indigenous Americans had been described as primitives who inhabited and fed off the forest, with little more impact on it than the squirrels themselves. To accommodate the Pilgrims' account of Thanksgiving, it was accepted that American Indians practiced limited, unobtrusive agriculture involving corn, beans, and squash.

We now know that many of the allegedly pristine landscapes of North and South America were actually artifacts, the result of enormous changes wrought by humans that started with the slaughter of megafauna. The first permanent Americans burned underbrush at least twice a year to make hunting easier. Most fires they set were low-intensity, meant to clear brambles and vermin, but they also selectively torched entire stands of trees to shape the forest into traps and funnels to corner wildlife.

The coast-to-Mississippi treetop traverse would have been possible only for birds. Not even flying squirrels could have managed it, because it took wings to cross large swathes where forest had been thinned to parkland or razed completely. By observing what grew after lightning opened clearings, paleo-Indians learned to create berry patches and herb-filled meadows to attract deer, quail, and turkeys. Finally, fire allowed them to do exactly what the Europeans and their descendants later came to do on such a grand scale: They farmed.

Yet there was one exception: New England, one of the first places where colonists arrived to stay, which may partly explain the familiar misconception of an entire virgin continent.

"There's now an understanding," says Harvard ecologist David Foster, "that precolonial eastern America had an agriculturally-based, maize-dependent large population with permanent villages and cleared fields. True. But that's not what we had up here."

It is a delicious September morning in deeply wooded central Massachusetts, just below the New Hampshire border. Foster has paused in a stand of tall white pines, which just a century earlier was a tilled wheat field. In their shady understory, little hardwoods are sprouting—maddening, he says, to timbermen who came after New England farmers had departed for points southwest and who thought they had a ready-made pine plantation.

"They spent decades of frustration trying to get white pine to succeed itself. They didn't get that when you cut down the forest, you expose a new forest that rooted in its shade. They never read Thoreau."

This is the Harvard Forest outside the hamlet of Petersham, established as a timber research station in 1907 but now a laboratory for studying what happens to land after humans no longer use it. David Foster, its director, has managed to spend much of his career in nature, not classrooms: at 50, he looks 10 years younger—fit and lean, the hair falling across his forehead still dark. He bounds over a brook that was widened for irrigation by one of the four generations of the family who farmed here. The ash trees along its banks are pioneers of the reborn forest. Like white pine, they don't regenerate well in their own shade, so in another century the small sugar maples beneath them will replace them. But this is already a forest by any definition: exhilarating smells, mushrooms popping through leaf litter, drops of green-gold sunlight, woodpeckers thrumming.

Even in the most industrialized part of a former farm, a forest resurges quickly here. A mossy millstone near a tumble of rocks that was once a chimney reveals where a farmer once ground hemlock and chestnut bark for tanning cowhides. The mill pond is now filled with dark sediment. Scattered firebricks, bits of metal and glass, are all that remain of the farmhouse. Its exposed cellar hole is a cushion of ferns. The stone walls that once separated open fields now thread between 100-foot conifers.

Over two centuries, European farmers and their descendants laid bare three-quarters of New England's forests, including this one. Three centuries more, and tree trunks may again be as wide as the monsters that early New Englanders turned into ship beams and churches—oaks 10 feet across, sycamores twice as thick, and 250-foot white pines. The early colonists found untouched, huge trees in New England, says Foster, because, unlike other parts of precolonial North America, this cold corner of the continent was sparsely populated.

"Humans were here. But the evidence shows low-density subsistence hunting and gathering. This isn't a landscape prone to burning. In all New England, there were maybe 25,000 people, not permanently in any one area. The postholes for structures are just two to four inches across. These hunter-gatherers could tear down and move a village overnight."

Unlike the center of the continent, says Foster, where large sedentary Native American communities filled the lower Mississippi Valley, New England didn't have corn until AD 1100. "The total accumulation of maize from New England archeological sites wouldn't fill up a coffee cup." Most settlements were in river valleys, where agriculture finally began, and on the coast, where maritime hunter-gatherers were sustained by immense stocks of herring, shad, clams, crabs, lobsters, and cod thick enough to catch by hand. Inland camps were mainly retreats from harsh coastal winters.

"The rest," says Foster, "was forest." It was a human-free wilderness, until Europeans named this land after their own ancestral home and proceeded to clear it. The timberlands the Pilgrims found were the ones that emerged in the aftermath of the last glaciers.

"Now we're getting that vegetation back. All the major tree species are returning."

So are animals. Some, like moose, have arrived on their own. Others, like beavers, were reintroduced and have taken off. In a world without humans to stop them, New England could return to what North America once looked like from Canada to northern Mexico: beaver dams spaced regularly on every stream, creating wetlands strung like fat pearls along their length, filled with ducks, muskrats, willets, and salamanders. One new addition to the ecosystem would be the coyote, currently trying to fill the empty wolf niche—though a new subspecies may be on the rise.

"The ones we see are substantially larger than western coyotes. Their skulls and jaws are bigger," says Foster, his long hands describing an impressive canine cranium. "They take larger prey than coyotes in the West, like deer. This probably isn't sudden adaptation. There's genetic evidence that western coyotes are migrating through Minnesota and up across Canada, interbreeding with wolves, then roaming here."

It's fortunate, he adds, that New England's farmers left before nonnative plants flooded America. Before exotic trees could spread across the land, native vegetation again had a roothold on their former farmlands. No chemicals had been spaded into their soils; no weeds, insects, or fungi here had ever been poisoned to help other things grow. It's the nearest thing to a baseline of how nature might reclaim cultivated land—against which to measure, for example, old England.

2. The Farm

Like most British trunk roads, the M1 motorway that runs north from London was built by Romans. In Hertfordshire, a jog at Hempstead leads to St. Alban's, once a substantial Roman town, and beyond that, to the village of Harpenden. From Roman times until the 20th century, when they became bedroom commutes to London, 30 miles away, St. Alban's was a center for rural commerce, and Harpenden was flat farmland, the conformity of its grain fields disrupted only by hedgerows.

Long before the Romans appeared in the first century AD, the dense forests of the British Isles began coming down. Humans first arrived 700,000 years ago, likely following herds of aurochs, the now-extinct wild Eurasian cattle, during glacial epochs when the English Channel was a land bridge, but their settlements were fleeting. According to the great British forest botanist Oliver Rackham, after the last ice age, southeastern

England was dominated by vast stands of lindens mixed with oaks, and by abundant hazels that probably reflect the appetites of Stone Age gatherers.

The landscape changed around 4,500 BC, because whoever crossed the water that by then separated England from the Continent brought crops and domestic animals. These immigrants, Rackham laments, "set about converting Britain and Ireland to an imitation of the dry open steppes of the Near East, in which agriculture had begun."

Today, less than 1/100 of Britain is original forest, and essentially none of Ireland. Most woodlands are clearly defined tracts, bearing evidence of centuries of careful human extraction by coppicing, which allowed stumps to regenerate for building supplies and fuel. They remained that way after Roman rule gave way to Saxon peasantry and serfdom, and into the Middle Ages.

At Harpenden, near a low stone circle and adjacent stem wall that are the remains of a Roman shrine, an estate was founded in the early 13th century. Rothamsted Manor, built of bricks and timbers and surrounded by a moat and 300 acres, changed hands five times over as many centuries, accruing more rooms until an eight-year-old boy named John Bennet Lawes inherited it in 1814.

Lawes went to Eton and then to Oxford, where he studied geology and chemistry, grew luxuriant muttonchops, but never took a degree. Instead, he returned to Rothamsted to make something of the estate his late father had left to seed. What he did with it ended up changing the course of agriculture and much of the surface of the Earth. How long those changes will persist, even after we're gone, is much debated by agro-industrialists and environmentalists. But with remarkable foresight, John Bennet Lawes himself has kindly left us many clues.

His story began with bones—although first, some would say, came chalk. Centuries of Hertfordshire farmers had dug the chalky remains of ancient sea creatures that underlie local clays to spread on their furrows, because it helped their turnips and grains. From Oxford lectures, Lawes knew that liming their fields didn't nourish plants so much as soften the soil's acidic resistance. But might anything actually feed crops?

A German chemist, Justus von Liebig, had recently noted that powdered bonemeal restored vigor to soil. Soaking it first in dilute sulfuric acid, he wrote, made it even more digestible. Lawes tried it on a turnip field. He was impressed.

Justus von Liebig is remembered as the father of the fertilizer industry, but he probably would have traded that honor for John Bennet Lawes's enormous success. It hadn't occurred to von Liebig to patent his process. After realizing what a bother it was for busy farmers to buy bones, boil them, grind them, then transport sulfuric acid from London gasworks to treat the crushed granules, and then mill the hardened result yet again, Lawes did so. Patent in hand, he built the world's first artificial fertilizer factory at Rothamsted in 1841. Soon he was selling "superphosphate" to all his neighbors.

His manure works—possibly at the insistence of his widowed mother, who still lived in the big brick manor—soon moved to larger quarters near Greenwich on the Thames. As the use of chemical soil additives spread, Lawes's factories multiplied, and his product line lengthened. It included not just pulverized bone and mineral phosphates, but two nitrogen fertilizers: sodium nitrate and ammonium sulfate (both later replaced by the ammonium nitrate commonly used today). Once again, the hapless von Liebig had identified nitrogen as a key component of amino and nucleic acids vital to plants, yet failed to exploit his discovery. While von Liebig published his findings, Lawes was patenting nitrate mixtures.

To learn which were most effective, in 1843 Lawes began a series of test plots still going today, which makes Rothamsted Research both the world's oldest agricultural station and also the site of the world's longest continual field experiments. Lawes and John Henry Gilbert, the chemist who became his partner of 60 years, earning the equal loathing of Justus von Liebig, began by planting two fields: one in white turnips, the other in wheat. They divided these into 24 strips, and applied a different treatment to each.

The combinations involved a lot, a little, or no nitrogen fertilizer; raw bonemeal, his patented superphosphates, or no phosphates at all; minerals such as potash, magnesium, potassium, sulfur, sodium; and both raw and cooked farmyard manure. Some strips were dressed with local chalk, some weren't. In subsequent years, some plots were rotated with barley, beans, oats, red clover, and potatoes. Some strips were periodically fallowed, some continually planted with the same crop. Some served as controls, with nothing added to them whatsoever.

By the 1850s, it was obvious that when both nitrogen and phosphate were applied, yields increased, and that trace minerals helped some crops and slowed others. With his partner, Gilbert, assiduously taking samples and recording results, Lawes was willing to test any theory—scientific, homespun, or wild—of what might help plants grow. According to his biographer, George Vaughn Dyke, these included trying superphosphate made from ivory dust, and slathering crops with honey. One experiment still running today involved no crops at all, only grass. An ancient sheep pasture just below Rothamsted Manor was divided into strips and treated with various inorganic nitrogen compounds and minerals. Later Lawes and Gilbert added fish meal and farm manure from animals fed different diets. In the 20th century, with increasing acid rain, the strips were further divided, with half receiving chalk to test growth under various pH levels.

From this pasture experiment, they noticed that although inorganic nitrogen fertilizer makes hay grow waist-high, biodiversity suffers. While 50 species of grass, weeds, legumes, and herbs might grow on unfertilized strips, adjacent plots dosed with nitrogen hold just two or three species. Since farmers don't want other seeds competing with the ones they've planted, they have no problem with this, but nature might.

Paradoxically, so did Lawes. By the 1870s, now wealthy, he sold his fertilizer businesses but continued his fascinating experiments. Among his concerns was how land could grow exhausted. His biographer quotes him as declaring that any farmer who thought he could "grow as fine crops by the aid of a few pounds of some chemical substances as by the same number of tons of farm-yard dung" was deluded. Lawes advised anyone planting vegetables and garden greens that, if it were him, he would "select a locality where I could obtain a large supply of yard manure at a cheap rate."

But in a rural landscape rushing to meet the dietary demands of a rapidly growing urban industrial society, farmers no longer had the luxury of raising enough dairy cows and pigs to produce the requisite tons of organic manure. Throughout densely populated late-19th-century Europe, farmers desperately sought food for their grain and vegetables. South Pacific islands were stripped of centuries of accumulated guano; stables were scoured for droppings; and even what was delicately called "night soil" was spread on fields. According to von Liebig, both horse and human bones from the Battle of Waterloo were ground and applied to crops.

As pressures on farmlands escalated in the 20th century, test plots at

Rothamsted Research were added for herbicides, pesticides, and municipal sewage sludge. The winding road to the old manor house is now lined with large laboratories for chemical ecology, insect molecular biology, and pesticide chemistry, owned by the agricultural trust that Lawes and Gilbert founded after both were knighted by Queen Victoria. Rothamsted Manor has become a dormitory for visiting researchers from around the world. Yet tucked behind all the gleaming facilities, in a 300-year-old barn with dusty windowpanes, is Rothamsted's most remarkable legacy.

It is an archive containing more than 160 years of human efforts to harness plants. The specimens, sealed in thousands of five-liter bottles, are of virtually everything. From each experimental strip, Gilbert and Lawes took samples of harvested grains, their stalks and leaves, and the soil where they grew. They saved each year's fertilizers, including manure. Later, their successors even bottled the municipal sewage sludge spread on Rothamsted test plots.

The bottles, stacked chronologically on 16-foot metal shelves, date back to the first wheat field in 1843. When mold developed in early samples, after 1865 they were stoppered with corks, then paraffin, and finally lead. During war years, when bottle supplies grew scarce, samples were sealed in tins that once held coffee, powdered milk, or syrup.

Thousands of researchers have mounted ladders to peruse the calligraphy on time-yellowed bottle labels—to extract, say, soil collected in Rothamsted's Geesecroft Field at a depth of nine inches in April 1871. Yet many bottles have never been opened: along with organic matter, they preserve the very air of their era. Were we to go suddenly, assuming no unprecedented seismic event dashes thousands of glass vessels to the floor, it's fair to surmise that this singular heritage would survive intact long beyond us. Within a century, of course, the durable slate-shingled roof would begin to yield to rain and vermin, and the smartest mice might learn that certain jars, when pushed to the concrete and shattered, contain still-edible food.

Supposing, however, that before such entropic vandalism occurs, the collection is discovered by visiting alien scientists who happen upon our now-quiet planet, bereft of voracious, but colorful, human life. Suppose they find the Rothamsted archive, its repository of more than 300,000 specimens still sealed in thick glass and tins. Clever enough to find their way to Earth, they would doubtless soon figure out that the graceful loops and symbols penned on the labels were a numbering system. Recognizing soil and

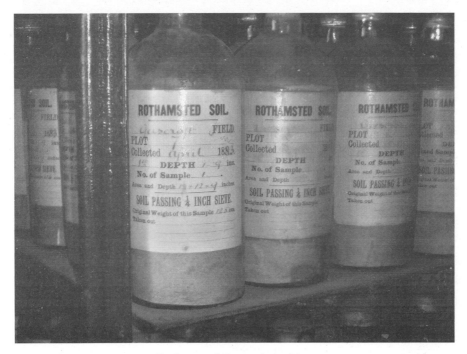

Rothamsted Research Archive.

PHOTO BY ALAN WEISMAN.

preserved plant matter, they might realize that they had the equivalent of a time-lapse record of the final century-and-a-half of human history.

If they began in the oldest jars, they would find relatively neutral soils that didn't stay that way for long as British industry redoubled. They would find the pH dropping farther into the acid end by the early 20th century, as the advent of electricity led to coal-fired power stations, which spread pollution beyond factory cities to the countryside. There would also be steadily increasing nitrogen and sulfur dioxide until the early 1980s, when improved smokestacks cut sulfur emissions so dramatically that the aliens might be puzzled to find samples spiked with powdered sulfur, which farmers had to start adding as fertilizer.

They might not recognize something that first appeared in Rothamsted's grassland plots in the early-1950s: traces of plutonium, a mineral that barely occurs in nature, let alone in Hertfordshire. Like grape vintages embodying annual weather, the fallout from tests in the Nevada desert, and later in Russia, marked Rothamsted's distant soils with their radioactive signature.

Uncorking the late 20th century, they would find that the bottles held other novel substances never before known on Earth (and, if they were lucky, not on their planet, either), such as polychlorinated biphenyls— PCBs—from the manufacture of plastics. To naked human eyes, the samples appear as innocent as comparable handfuls of dirt in specimen bottles from 100 years earlier. Alien vision, however, might discern menaces we only see with devices like gas chromatographs and laser spectrometers.

If so, they might glimpse the sharp fluorescent signature of polyaromatic hydrocarbons (PAHs). They might be astonished at how PAHs and dioxins, two substances emitted naturally by volcanoes and forest fires, suddenly leaped from background levels into center-stage chemical prominence in soil and crops as the decades advanced.

If they were carbon-based life-forms like us, they might leap themselves, or at least back away, because both PAHs and dioxins can be lethal to nervous systems and other organs. PAHs were buoyed into the 20th century aboard clouds of exhaust from automobiles and coal-fired power plants; they're also in the pungent odor of fresh asphalt. At Rothamsted, as at farms everywhere, they were introduced deliberately, in herbicides and pesticides.

Dioxins, however, were unintended: they're by-products formed when hydrocarbons combine with chlorine, with tenacious, disastrous results. Besides their role as sex-changing endocrine disruptors, their most infamous application before being banned was in Agent Orange, a defoliant that laid bare entire Vietnamese rain forests so that insurgents would have nowhere to hide. From 1964 to 1971, the United States doused Vietnam with 12 million gallons of Agent Orange. Four decades later, heavily dosed forests still haven't grown back. In their place is a grass species, cogon, called one of the world's worst weeds. Burned off constantly, it keeps springing back, overwhelming attempts to supplant it with bamboo, pineapple, bananas, or teak.

Dioxins concentrate in sediments, and thus show up in Rothamsted's sewage sludge samples. (Municipal sludge, since 1990 deemed too toxic to dump into the North Sea, is instead spread as fertilizer on European farmlands—except in Holland. Since the 1990s, the Netherlands has not only offered incentives that practically equate organic farming with patriotism, but has also struggled to convince its EU partners that everything applied to the land ends up in the sea anyway.)

Will the future visitors who discover Rothamsted's extraordinary archive wonder if we were trying to kill ourselves? They might find hope in the fact that, beginning in the 1970s, lead deposition in soil waned significantly. But at the same time, the presence of other metals was increasing. Especially in preserved sludge, they would find all the nasty heavies: lead, cadmium, copper, mercury, nickel, cobalt, vanadium, and arsenic, and also lighter ones like zinc and aluminum.

3. The Chemistry

Dr. Steven McGrath hunches over his corner computer, deep-set eyes beneath his gleaming pate crinkling through rectangular reading lenses at a map of Britain and a chart color-coded with things that on an ideal planet— or one that gets the chance to start over—wouldn't show up in plants that animals like to eat. He points at something yellow.

"This, for instance, is the net accumulation of zinc since 1843. No one else can see these trends because our samples," he adds, his shirtfront slightly inflating, "are the longest test archive in the world."

From sealed samples of a winter-wheat field called Broadbalk, one of Rothamsted's oldest, they know that the original 35 parts per million of zinc present in the soil have nearly doubled. "That's coming from the atmosphere, because our control plots have nothing added—no fertilizers, no manure or sludge. Yet the concentration is up 25 ppm."

The test farm plots, however, which also originally had 35 ppm of zinc, now are at 91 ppm. To the 25 ppm from airborne industrial fallout, something is adding another 31 ppm.

"Farmyard manure. Cows and sheep get zinc and copper in their animal feed to keep them healthy. Over 160 years, it's nearly doubled the zinc in the soil."

If humans disappeared, so would zinc-laced smoke from factories, and no one would be feeding mineral supplements to livestock. Yet McGrath expects that, even in a world without people, metals we put into the ground will be around a long time. How long before rain leaches them out, returning soils to a preindustrial state, would depend, McGrath says, on their composition.

"Clays will hang onto them up to seven times as long as sandy soils, because they don't drain as freely." Peat, also poorly drained, can retain lead, sulfur, and organochloride pollutants like dioxin even longer than clay. McGrath's maps show hot clusters on peat-covered hilltops on the English and Scottish moors.

Even sandy soils can bind nasty heavy metals when municipal sludge is mixed into them. In sludged earth, leaching of metals drops as chemical bonds form; extraction is mainly via roots. Using archived samples of Rothamsted carrots, beets, potatoes, leeks, and various grains treated since 1942 with West Middlesex municipal sludge, McGrath has calculated how long metals were added to such soil will stay there—assuming crops are still being harvested.

From a file drawer, he produces a table that gives the bad news. "With no leaching, I figure zinc lasts 3,700 years."

That's how long it took humans to get from the Bronze Age to today. Compared to the time other metallic pollutants would linger, that turns out to be short. Cadmium, he says, an impurity in artificial fertilizer, will cling twice as long: 7,500 years, or the same amount of time that's passed since humans began irrigating Mesopotamia and the Nile Valley.

It gets worse. "Heavier metals like lead and chromium tend not to be taken up as easily by crops, and not to be leached. They simply bind." Lead, the one with which we've most recklessly laced our topsoil, will take nearly 10 times as long as zinc to disappear—the next 35,000 years. Thirty-five thousand years ago was a couple of ice ages back.

For unclear chemical reasons, chromium is the most stubborn of all: Mc-Grath estimates 70,000 years. Toxic in mucous membranes or if swallowed, chromium leaks into our lives mainly from tanning industries. Smaller amounts are chipped from aging chrome-plated sink taps, brake linings, and catalytic converters. But compared to lead, chromium is a minor concern.

Humans discovered lead early, but only recently realized how it afflicts nervous systems, learning development, hearing, and general brain function. It also causes kidney disease and cancer. In Britain, Romans smelted lead from mountain-ore veins to make pipes and chalices—poisonous choices suspected to have left many people dead or demented. The use of lead plumbing continued through the Industrial Revolution—Rothamsted Manor's historic storm drains bearing ornate family crests are still lead.

But old plumbing and smelting add just a few percentage points of lead to our ecosystem. Will our visitors who arrive sometime in the next 35,000 years deduce that vehicle fuel, industrial exhaust, and coal-fired power plants spewed the lead they detect everywhere? Since no one will harvest whatever grows in metal-saturated fields after we're gone, Mc-Grath guesses that plants will keep taking it up, then putting it back as they die and decay, in a continuous loop.

Through genetic tinkering, both tobacco and a flower called mouse-ear cress have been modified to suck up and exhale one of the most dreaded heavy-metallic toxins of all, mercury. Unfortunately, plants don't redeposit metals deep in the Earth where we originally dug them. Breathe away mercury, and it rains down elsewhere. There's an analogy, Steve McGrath says, to what happens with PCBs—the polychlorinated biphenyls once used in plastics, pesticides, solvents, photocopying paper, and hydraulic fluids. Invented in 1930, they were outlawed in 1977 because they disrupt immune systems, motor skills, and memory, and play roulette with gender.

Initially, banning PCBs seemed to have worked: Rothamsted's archive clearly shows their presence in soils dropping through the 1980s and 1990s until, by the new millennium, they practically reached preindustrial levels. Unfortunately, it turns out that they merely wafted away from the temperate regions where they were used, then sunk like chemical stones when they hit cold air masses in the Arctic and Antarctic.

The result is elevated PCBs in the breast milk of Inuit and Laplander mothers, and in the fat tissues of seals and fish. Along with other pole-bound POPs—"persistent organic pollutants"—such as polybrominated diphenyl flame retardants, or PBDEs, PCBs are the suspected culprits for growing numbers of hermaphroditic polar bears. Neither PCBs nor PBDEs existed until humans conjured them up. They consist of hydrocarbons wedded to highly reactive elements known as halogens, like chlorine or bromine.

The acronym POPs sounds regrettably light-hearted, because these substances are all business, designed to be extremely stable. PCBs were the fluids that kept on lubricating; PBDEs the insulator that kept plastic from melting; DDT the pesticide that kept on killing. As such, they are difficult to destroy; some, like PCBs, show little or no sign of biodegrading.

As the flora of the future keep recycling our metals and POPs for the next several thousand years, some will prove tolerant; some will adapt to a metallic flavor in their soil, as the foliage growing around Yellowstone geysers has done (albeit over a few million years). Others, however—like some of us humans—will die from lead or selenium or mercury poisoning. Some of those that succumb will be weak members of a species that will then grow stronger as it selects for yet a new trait, such as mercury or DDT tolerance. And some species will be selected out entirely, and go extinct.

After we're gone, the lasting effects of all the fertilizers we've spread on furrows since John Lawes began hawking them will vary. Some soils, their pH depressed from years of nitrates diluting to nitric acid, may recover in decades. Others, such as those in which naturally occurring aluminum concentrates to toxic proportions, won't grow anything until leaf litter and microbes make soil all over again.

The worst impact of phosphates and nitrates, however, isn't in fields, but where they drain. Even more than a thousand miles downstream, lakes and river deltas suffocate beneath over-fertilized aquatic weeds. Mere pond scum morphs into algae blooms weighing tons, which suck so much oxygen from freshwater that everything swimming in it dies. When the algae collapse, their decay escalates the process. Crystalline lagoons turn to sulfurous mudholes; estuaries of eutrophic rivers balloon into gigantic dead zones. The one spreading into the Gulf of Mexico at the mouth of the Mississippi, charged with fertilizer-soaked sediments all the way from Minnesota, is now bigger than New Jersey.

In a world without humans, a screeching halt to all artificial farmland fertilization would take instant, enormous chemical pressure off the richest biotic zones on Earth—the areas where big rivers bearing huge natural nutrient loads meet the seas. Within a single growing season, lifeless plumes from the Mississippi to the Sacramento Delta, to the Mekong, Yangtze, Orinoco, and the Nile, would begin to shrink. Repeated flushings of a chemical toilet will steadily clarify the waters. A Mississippi Delta fisherman who awakened from the dead after only a decade would be amazed at what he'd find.

4. The Genes

Since the mid-1990s, humans have taken an unprecedented step in Earthly annals by introducing not just exotic flora or fauna from one ecosystem into another, but actually inserting exotic genes into the operating systems of individual plants and animals, where they're intended to do exactly the same thing: copy themselves, over and over.

Initially, GMOs—genetically modified organisms—were conceived to make crops produce their own insecticides or vaccines, or to make them invulnerable to chemicals designed to kill weeds competing for their furrows, or to make them—and animals as well—more marketable. Such product improvement has extended the shelf life of tomatoes; spliced DNA from Arctic Ocean fish into farmed salmon so that they churn out growth hormones year-round; induced cows to give more milk; beautified the grain in commercial pine; and imbued zebra fish with jellyfish fluorescence to spawn glow-in-the-dark aquarium pets.

Growing more ambitious, we've coaxed plants that we feed to animals to also deliver antibiotics. Soybeans, wheat, rice, safflower, canola rapeseed, alfalfa, and sugarcane are being genetically hot-rodded to produce everything from blood thinners to cancer drugs to plastics. We've even bio-enhanced health food to produce supplements like beta carotene or gingko biloba. We can grow wheat that tolerates salt and timber that resists drought, and we can make various crops either more or less fertile, depending on which is desired.

Appalled critics include the U.S.-based Union of Concerned Scientists, and approximately half of Western Europe's provinces and counties, including much of the United Kingdom. Among their fears is what we might do to the future, should some new life-form proliferate like kudzu. Crops such as Monsanto's suite of "Roundup Ready" corn, soy, and canola—molecularly armored to shrug off that company's flagship herbicide while everything else nearby dies—are doubly dangerous, they insist.

For one, they say, sustained use of Roundup—a trade name for glyphosate—on weeds has simply selected for Roundup-resistant strains of weeds, which then drive farmers to use additional herbicides. Second, many crops broadcast pollen to propagate. Studies in Mexico that show bio-tinkered corn invading neighboring fields and cross-pollinating natural strains have provoked denials and pressure on university researchers by the food industry, which underwrites much of the funding for expensive genetic studies.

Modified genes from commercially bred bentgrass, a turf used on golf courses, have been confirmed in native Oregon grasses, miles from the source. Assurances from the aquaculture industry that genetically supercharged salmon won't breed with wild North American stock, because

they're raised in cages, are belied by thriving salmon populations in estuaries in Chile—a country that had no salmon until breeders were imported from Norway.

Not even supercomputers can predict how man-made genes already
loosed upon the Earth will react in a near infinity of possible eco-niches.
Some will be roundly trounced by competition toughened over eons by
evolution. It's a fair bet, though, that others will pounce on an opportunity
to adapt, and evolve themselves.

5. Beyond the Farm

Rothamsted research scientist Paul Poulton stands in November drizzle,
knee-deep in holly, surrounded by what will be around after human cultivation ceases. Born just a few miles up the road, lanky Paul Poulton is as
rooted to this land as any crop. He started working here right out of
school, and now his hair has whitened. For more than 30 years, he's
tended experiments that began before he was born. He'd like to think they
will continue on long after he himself turns to bone dust and compost.
But one day, he knows, the wild green profusion beneath his muddy irrigation boots will be the only Rothamsted experiment that will still matter.

It is also the only one that has required no management. In 1882, it occurred to Lawes and Gilbert to fence off a half-acre of Broadbalk—the
winter-wheat field that had variously received inorganic phosphate, nitrogen,
potassium, magnesium, and sodium—and leave the grain unharvested, just
to see what would happen. The following year, a new crop of self-seeded
wheat appeared. The year after that, the same thing occurred, though by now
invading hogweed and creeping woundwort were competing for the soil.

By 1886, only three stunted, barely recognizable wheat stalks germinated. A serious incursion of bentgrass had also appeared, as well as a scattering of yellow wildflowers, including orchid-like meadow peas. The next
year, wheat—that robust Middle Eastern cereal grown here even before the
Romans arrived—had been entirely vanquished by these returning natives.

Around that time, Lawes and Gilbert abandoned Geescroft, a parcel
about half a mile away, consisting of slightly more than three acres. From
the 1840s to the 1870s, it had been planted in beans, but after 30 years, it
was clear that even with chemical boosts, growing beans continuously

without rotation was a failure. For a few seasons, Geescroft was seeded in red clover. Then, like Broadbalk, it was fenced off to fend for itself.

For at least two centuries before Rothamsted's experiments began, Broadbalk had received dressings of local chalk, but low-lying Geescroft, hard to cultivate without digging drainage, apparently hadn't. In the decades following abandonment, Geesecroft turned increasingly acidic. At Broadbalk, which was buffered by years of heavy liming, pH had barely lowered. Complex plants like chickweed and stinging nettle were showing up there, and within 10 years filbert, hawthorn, ash, and oak seedlings were establishing themselves.

Geescroft, however, remained mainly a prairie of cocksfoot, red and meadow fescue, bentgrass, and tufted hair grass. Thirty years would pass before woody species began shading its open spaces. Broadbalk, meanwhile, grew tall and dense. By 1915 it added 10 more tree types, including field maple and elm, plus thickets of blackberry and a dark green carpet of English ivy.

As the 20th century progressed, the two parcels continued their separate metamorphoses from farmland to woodland, the differences between them amplifying as they matured, echoing their distinct agricultural histories. They became known as the Broadbalk and Geescroft Wildernesses—a seemingly pretentious term for land totaling less than four acres, yet perhaps fitting in a country with less than 1 percent of its original forests remaining.

In 1938, willows sprouted around Broadbalk, but later they were replaced by gooseberry and English yew. "Here in Geescroft," says Paul Poulton, unsnagging his rain parka from a bush bright with berries, "there was none of this. Suddenly, 40 years ago, holly started coming in. Now we're overgrown. No idea why."

Some of the holly bushes are the size of trees. Unlike Broadbalk, where ivy swirls up the trunk of every hawthorn and flows over the forest floor, there is no ground cover, save for brambles. The grasses and herbaceous weeds that first colonized Geescroft's fallowed field are completely gone, shaded out by oaks, which prefer acidic soil. Due to overplanting of nitrogen-fixing legumes, and also to nitrogen fertilizers and decades of acid rain, Geesecroft is a classic example of exhausted soil, acidified and leached, with only a few species predominating.

Even so, a forest of mainly oak, brambles, and holly is not a barren place. It is life that, in time, will beget more.

Broadbalk Wheatfield and "Wilderness." (Trees, upper left.)
© ROTHAMSTED RESEARCH LTD 2003.

The difference at Broadbalk—which has just one oak—is two centuries of chalk lime, which retains phosphates. "But eventually," says Poulton, "it will wash out." When it does, there will be no recovery, because once the calcium buffer is gone, it can't return naturally unless men with shovels return to spread it. "Someday," he says, almost in a whisper, his thin face scanning his life's work, "all this farmland will go back to woody scrub. All the grass will disappear."

Without us, it will take no more than a century. Rinsed of its lime, Broadbalk Wilderness will be Geescroft revisited. Like arboreal Adams and Eves, their seeds will cross on the winds until these two remnant woodlands merge and spread, taking all the former fields of Rothamsted back to their unfarmed origins.

In the mid-20th century, the length of commercialized wheat stalks shortened nearly by half even as the number of grains they bore multiplied. They were engineered crops, developed during the so-called Green Revolution to eliminate world hunger. Their phenomenal yields fed millions who otherwise might not have eaten, and thus also contributed to expanding the populations of countries like India and Mexico. Designed through forced crossbreeding and random mixes of amino acids—tricks that preceded gene splicing—their success and survival depend on calibrated cocktails of fertilizers, herbicides, and pesticides to protect these laboratory-bred life-forms from perils that lurk outside, in reality.

In a world without people, none will last in the wild even the four years during which wheat hung on in the Broadbalk Wilderness after Lawes and Gilbert abandoned it to the elements. Some are sterile hybrids, or they spawn offspring so defective that farmers must purchase new seed each year—a boon for seed companies. The fields where they are destined to die out, which are now most of the grain fields in the world, will be left deeply soured by nitrogen and sulfur, and will remain badly leached and acidic until new soil is built. That will require decades of acid-tolerant trees rooting and growing, then hundreds of years more of leaf litter and decaying wood broken down and excreted as humus by microbes that can tolerate the thin legacy of industrial agriculture.

Beneath these soils, and periodically disinterred by ambitious root systems, will lie three centuries' worth of various heavy metals and an alphabet soup of POPs, substances truly new under the sun and soil. Some engineered compounds like PAHs, too heavy to blow away to the Arctic, may end up molecularly bound in soil pores too tiny for digesting microbes to enter, and remain there forever.

∼ૐ∾

IN 1996, LONDON journalist Laura Spinney, writing in *New Scientist Magazine*, envisioned her city abandoned 250 years hence, turned back into the swamp it once was. The liberated Thames wandered among the waterlogged foundations of fallen buildings, Canary Wharf Tower having collapsed under an unbearable tonnage of dripping ivy. The following year, Ronald Wright's novel *A Scientific Romance* jumped 250 years more, and imagined the same river lined with palms, flowing transparently past Canvey Island into a sweltering mangrove estuary, where it joined a warm North Sea.

Like the entire Earth, the posthuman fate of Britain teeters on the balance of these two visions: a return to temperate foliage, or a lurch into a tropical, super-heated future—or, ironically, into a semblance of something last seen in England's southwestern moors, where Conan Doyle's Baskerville hound once wailed into chill mist.

Dartmoor, the highest point in southern England, resembles a 900-square-mile baldpate with occasional massive chunks of fractured granite poking through, fringed by farms and patches of woods that exploded from old boundary hedgerows. It formed at the end of the Carboniferous Age, when most of Britain lay submerged, with sea creatures dropping shells on what became its buried chalk. Beneath that was granite, which 300 million years ago bulged with underlying magma into a dome-shaped island—which it may be again if seas rise as high as some fear.

Several ice ages froze enough of the planet's water solid to drop ocean levels and allow today's world to take shape. The last of these sent a mile-high ice sheet right down the Prime Meridian. Where it stopped is where Dartmoor begins. Atop its granite hilltops, known as tors, are remnants from those times that may be portents of what awaits if yet a third climatic alternative proves to be the British Isles' destiny.

That fate could occur if meltwater from Greenland's ice cap shuts down, or actually reverses, the oceanic conveyor atop which rides the Gulf Stream, which currently keeps Britain far warmer than Hudson's Bay, at the same latitude. Since that much-debated event would be the direct result of rising global temperatures, probably no ice sheet will form—but permafrost and tundra could.

That happened at Dartmoor 12,700 years ago, the last time the global circulation system nearly slowed to a halt: no ice, but rock-hard ground. What followed is not only instructive, as it shows what the United Kingdom might resemble in coming years, but also hopeful, because these things, too, will pass.

The deep freeze lasted 1,300 years. During that time, water trapped in fissures in Dartmoor's granite dome bedrock froze, cracking apart huge rocks below the surface. Then the Pleistocene ended. The permafrost thawed; its runoff exposed the shattered granite that became Dartmoor's tors, and the moor bloomed. Across the land bridge that for another 2,000 years connected England to the rest of Europe, pine moved in, then birch, then oak. Deer, bears, beavers, badgers, horses, rabbits, red squirrels, and aurochs crossed with them. So did a few significant predators: foxes, wolves, and the ancestors of many of today's Britons.

As in America, and Australia long before, they used fire to clear trees, making it easier to find game. Except for the highest tors, the barren Dartmoor prized by local environmental groups is another human artifact. It is a former forest repeatedly burned, then waterlogged by more than 100 inches of annual rainfall into a blanket of peat where trees no longer grow. Only charcoal remnants in peat cores attest that once they did.

The artifact was shaped further as humans pushed hunks of granite into circles that became foundations for their huts. They spread them into long, low unmortared stone reaves that crossed and hatched the landscape, and remain vivid even today.

The reaves divided the land into pastures for cows, sheep, and Dartmoor's famous hardy ponies. Recent attempts to emulate Scotland's picturesque heaths by removing livestock proved futile, as bracken and prickly gorse appeared rather than purple heather. But gorse befits a former tundra, whose frozen surfaces melt to spongy peat familiar to anyone who walks these moors. Tundra this may be again, whether humans are here or not.

Elsewhere on Earth, on former croplands that humans tended for millennia, warming trends will create variations of today's Amazon. Trees may cover them with vast canopies, but the soils will remember us. In the Amazon itself, charcoal that permeates frequent deposits of rich black soil called *terra preta* suggests that, thousands of years ago, paleo-humans cultivated wide swathes of what we think of today as jungle primeval. Slowly charring rather than burning trees, they ensured that much of their nourishing carbon was not expelled into the atmosphere but was instead retained, along with nitrogen, phosphorus, calcium, and sulfur nutrients—all packaged in easily digested organic matter.

This process has been described by Johannes Lehmann, the latest of a lineage of Cornell University soil scientists who have studied *terra preta* nearly as long as the heirs to Rothamsted founder John Lawes have experimented with fertilizer. The charcoal-enriched soil, despite incessant use, never gets depleted. Witness the lush Amazon itself: Lehmann and others believe that it sustained large pre-Columbian populations, until European diseases reduced them to scattered tribes who now live off nut groves planted by their ancestors. The unbroken Amazon we see today, the world's largest forest, rushed back so quickly across rich *terra preta* that European colonists never realized it was gone.

"Producing and applying bio-char," writes Lehmann, "would not only dramatically improve soil and increase crop production, but also could provide a novel approach to establishing a significant, long-term sink for atmospheric carbon dioxide."

In the 1960s, British atmospheric scientist, chemist, and marine biologist James Lovelock proposed his Gaia hypothesis, which describes the Earth as behaving like a super-organism, its soil, atmosphere, and oceans composing a circulatory system regulated by its resident flora and fauna. He now fears that the living planet is suffering a high fever, and that we are the virus. He suggests we compile a user's manual of vital human knowledge (on durable paper, he adds) for survivors who may sit out the next millennium huddled in the polar regions, the last habitable places in a super-heated world, until the ocean recycles enough carbon to restore a semblance of equilibrium.

If we do so, the wisdom of those nameless Amazonian farmers should be inscribed and underlined so that we might attempt agriculture a little differently next time around. (There may be a chance: Norway is now archiving examples of the world's crop seed varieties on an Arctic island, in hopes they may survive untold calamities elsewhere.)

If not, and if no humans return to till the soil or husband the animals, forests will take over. Rangelands that receive good rainfall will welcome new grazers—or old ones, as some new incarnation of *Proboscidae* and sloths replenish the Earth. Other places, however, less blessed, will have parched into new Saharas. The American Southwest, for instance: waist-high in grasses until 1880, when their cattle population of a half-million suddenly sextupled, New Mexico and Arizona now face unprecedented drought, with much of their water-retaining capacity lost. They may have to wait.

Still, the Sahara itself was once covered with rivers and ponds. With patience—though not, unfortunately, human patience—it will be again.

PART III

༺༻

CHAPTER 12

The Fate of Ancient and Modern Wonders of the World

Between global warming and ocean-conveyor cooling, if whichever dominates is partly muted by the other, as some models propose, Europe's meticulous mechanized farmlands would, without humans, fill with brome and fescue grass, lupine, plumed thistle, flowering rapeseed, and wild mustard. Within a few decades oak shoots would sprout from the acidic former fields of wheat, rye, and barley. Boars, hedgehogs, lynx, bison, and beaver would spread, with wolves moving up from Romania and, if Europe is cooler, reindeer coming down from Norway.

The British Isles would be somewhat biologically marooned, as rising seas batter the already-receding chalk cliffs of Dover and widen the 21-mile gap that separates England from France. Dwarf elephants and hippopotamuses once may have swum almost double that distance to reach Cyprus, so presumably something might try. Caribou, buoyed by their insulating hollow hair, cross northern Canadian lakes, so their reindeer siblings might just make it to England.

Should some impetuous animal attempt the journey via Chunnel—the English Channel Tunnel, *Le Tunnel sous la Manche*—after human traffic ceases, it might actually make it. Even without maintenance, the Chunnel wouldn't quickly flood like many of the world's subways, because it was dug within a single geologic layer, a bed of chalk marl with minimal filtration.

Whether an animal would actually try is another matter. All three

Chunnel tubes—one each for westbound and eastbound trains, and a parallel central corridor to service them—are swaddled in concrete. For 35 miles there would be no food or water—just pitch darkness. Still, it's not impossible that some continental species might recolonize Britain that way: The capacity of organisms to ensconce themselves in the world's most inhospitable places—from lichens on Antarctic glaciers to sea worms in 176°F sea vents—may symbolize the meaning of life itself. Surely, as small, curious creatures like voles or the inevitable Norway rats slither down the Chunnel, some brash young wolf will follow their scent.

The Chunnel is a true wonder of our times, and, at a cost of $21 billion, also the most expensive construction project ever conceived until China began damming several rivers at once. Protected by its buried bed of marl, it has one of the best chances of any human artifact to last millions of years, until continental drift finally pulls it apart or scrunches it like an accordion.

While still intact, however, it may not remain functional. Its two terminals are just a few miles from their respective coasts. There's little chance that the Folkestone, England, entrance, nearly 200 feet above current sea level, could be breached: the chalk cliffs that separate it from the English Channel would have to erode significantly. Far more likely is that ascending waters could enter the Coquelles, France, terminal, only about 16 feet above sea level on the Calais plain. If so, the Chunnel would not completely flood: the marl stratum it follows makes a mid-channel dip and then rises, so water would seek the lowest levels, leaving part of the chambers clear.

Clear, but useless, even to daring migrating creatures. But when $21 billion was spent to create one of engineering's greatest wonders, no one imagined that the oceans might rise up against us.

Nor did the proud builders of the ancient world, which had seven wonders, dream that in a span far shorter than eternity only one of them—Egypt's Khufu pyramid—would remain. Like old-growth forest whose lofty treetops eventually collapse, Khufu has shrunk some 30 feet over the past 4,500 years. At first, that was no gradual loss—its marble shell was cannibalized during the Middle Ages by conquering Arabs to build Cairo. The exposed limestone is now dissolving like any other hill, and in a million more years should not look very pyramidal at all.

The other six were of even more mortal stuff: a huge wooden idol of Zeus plated in ivory and gold, which fell apart during an attempt to move it; a hanging garden, of which no trace remains among the ruins of its Babylonian palace 30 miles south of Baghdad; a colossal bronze statue on Rhodes that collapsed under its own weight in an earthquake and was later sold for scrap; and three marble structures—a Greek temple that crumbled in a fire, a Persian mausoleum razed by Crusaders, and a lighthouse marking Alexandria's harbor, which was felled by earthquake as well.

What made them qualify as wonders was sometimes stirring beauty, as in the case of the Temple of Artemis in Greece, but more often it was simply massive scale. Human creation writ very large often overwhelms us into submission. Less ancient, but most imposing of all, is a construction project that spanned 2,000 years, three ruling dynasties, and 4,000 miles, resulting in a rampart so monumental that it achieved the status not just of landmark, but land*form*. The Great Wall of China is so staggering that it was widely, although erroneously, believed visible from outer space, serving notice even to would-be attackers from other worlds that this property was defended.

Yet, like any other ripple in the Earth's crust, the Great Wall is not immortal, and far less so than most geologic versions. A pastiche of rammed earth, stones, fired brick, timbers, and even glutinous rice used as mortar paste, without human maintenance it is defenseless against tree roots and water—and the highly acidic rain produced by an industrializing Chinese society isn't helping. Yet without that society, it will steadily melt away until just the stones remain.

Walling the Earth from the Yellow Sea all the way across Inner Mongolia is impressive, but for grand public works, few have matched a modern wonder whose construction began in 1903, the same year that New York inaugurated its subway. It was no less than the human race defying plate tectonics by tearing apart two continents that floated together 3 million years earlier. Nothing like the Panama Canal had ever been attempted before, and little has come close to it since.

Although the Suez Canal had already severed Africa from Asia three decades earlier, that was a comparatively simple, sea-level surgical stroke across an empty, disease-free sand desert with no hills. The French company that dug it went next to the 56-mile-wide isthmus between the Americas, smugly intending to do the same. Disastrously, they underestimated dense jungle steeped in malaria and yellow fever, rivers fed by prodigious rainfall, and a continental divide whose lowest pass was still 270 feet above the sea. Before they were one-third of the way through, they suffered not only a bankruptcy that rocked France, but also the deaths of 22,000 workers.

Nine years later, in 1898, a highly ambitious Assistant Secretary of the Navy named Theodore Roosevelt found a pretext, based on an explosion (probably due to a faulty boiler) that sank a U.S. ship in Havana Harbor, to oust Spain from the Caribbean. The Spanish-American War was intended to liberate both Cuba and Puerto Rico, but, to the great surprise of Puerto Ricans, the United States annexed their island. To Roosevelt, it was perfectly positioned as a coaling station for the still-nonexistent canal that would eliminate the need for ships sailing between the Atlantic and the Pacific to travel down the length of South America and up again.

Roosevelt chose Panama over Nicaragua, whose eponymous navigable lake, which would have saved considerable digging, lay among active volcanoes. At the time, the isthmus was part of Colombia, although Panamanians had tried three times to bolt from distant Bogota's fitful rule. When Colombia objected to the U.S. offer of just $10 million for sovereignty over a 6-mile-wide zone bordering the proposed canal, President Roosevelt sent a gunboat to help Panamanian rebels finally succeed. A day later, he betrayed them by recognizing as Panama's first ambassador to the United States a French engineer from France's defunct canal-digging company, who, at considerable personal profit, immediately affirmed a treaty agreeing to U.S. terms.

That sealed the United States' reputation in Latin America as piratical gringo imperialists, and produced—11 years and 5,000 more deaths later—the most stunning engineering feat yet in human history. More than a century has passed and it is still among the greatest of all time. Besides reconfiguring continental landmasses and communication between two oceans, the Panama Canal also significantly shifted the economic center of the world to the United States.

Something that substantial and literally earth-moving seems destined to last for the ages. But in a world without us, how long would it take nature to rejoin what man split asunder in Panama?

⤜⤝

"THE PANAMA CANAL," says Abdiel Pérez," is like a wound that humans inflicted on the Earth—one that nature is trying to heal."

As superintendent of the locks on the Canal's Atlantic side, Pérez—along with 5 percent of all planetary commerce—depends on a handful of hydrologists and engineers charged with keeping that wound open. A square-jawed, soft-voiced electrical and mechanical engineer, Pérez began here in the 1980s as an apprentice machinist while studying at the University of Panama. Daily, he feels humbled to be entrusted with one of the most revolutionary pieces of machinery on Earth.

"Portland cement was a novelty. This is where it was tried out. Reinforced concrete wasn't invented yet. All the walls of the locks are oversized like a pyramid. Their only reinforcement is gravity."

He stands alongside what is essentially a huge concrete box, into which an orange Chinese freighter bound for the East Coast of the United States, stacked seven stories high with containers, has just been guided. The lock is 110 feet wide. The ship, as long as three football fields, has exactly two feet of clearance on each side as two electric railway engines, called *mules,* tug it through the glove-tight locks.

"Electricity was also new. New York had barely installed the first generating plant. But the Canal builders decided to use electricity, not steam engines."

Once the ship is inside, water is piped into the lock to raise it 28 feet, which takes ten minutes. On the lock's opposite end awaits Lake Gatún, for a half-century the biggest artificial lake in the world. Creating it drowned an entire mahogany forest, but prevented a repeat of the French debacle, which resulted from the fatal decision to try digging another sea-level canal like Suez. Besides entailing removal of a large chunk of the continental divide, there was also the matter of the Río Chagres, a rain-gorged river that, as it plunged from jungle highlands to the sea, smacked into the middle of the canal's route. During Panama's eight-month rainy season, the Chagres carried enough silt to plug a narrow man-made channel in mere days, if not hours.

The Americans' solution was to build an aquatic staircase fashioned by three locks on either end, rising in watery steps to a lake formed by the dammed Chagres in the middle—a liquid bridge over which boats could float across the hills that the French failed to cut through. The locks use 52,000 gallons of water to lift every ship that passes through—freshwater fed by gravity from the trapped river, which drains to the sea as each vessel exits. Although gravity is always available, the electricity that opens and closes the doors of each lock depends on human operators who maintain hydroelectric generators that also tap the Chagres.

There are also auxiliary steam power and a diesel plant, but, says Pérez, "without people, the electricity wouldn't last a day. Someone in control must decide where the power's coming from, whether to open or close turbines, et cetera. With no human in the system, it doesn't work."

What particularly wouldn't work are the 7-foot-thick hollow, floating steel doors, 80 feet high and 65 feet wide. Each lock has a double set as backup, pivoting on plastic bearings that, during the 1980s, replaced the original brass hinges that corroded every few decades. What if power were cut, and the doors opened and stayed that way?

"Then it's all over. The highest lock is 137 feet above sea level. Even if they were left closed, once their seals went, so would the water." The seals are steel plates overlapping each door's leading edge, which need replacing every 15 to 20 years. Pérez glances up as the shadow of a frigate bird speeds past, then resumes watching the double doors close behind the departing Chinese freighter.

"The whole lake could empty through the locks."

Gatún Lake sprawls over what was once the course of the Río Chagres as it emptied into the Caribbean. Reaching it from the Pacific side required cutting through the 12 miles of terrestrial spine that bisects Panama lengthwise at La Culebra, the lowest saddle in the continental divide. Slicing through that much soil, iron oxide, clay, and basalt would have been daunting anywhere, but even after the French disaster, no one really understood how truly unstable the waterlogged Panamanian earth was.

The Culebra Cut initially was to be 300 feet wide. As one gigantic mudslide after another undid months of digging, sometimes burying box-cars and steam shovels as it refilled the trench, engineers had to keep widening the slope. In the end, the mountain range that runs from Alaska to Tierra del Fuego was separated in Panama by a man-made valley, its gap about six times as broad as its floor. To dig it required the labor of 6,000 men every day for seven years. The 100 million–plus cubic yards of dirt they moved, if compacted together, would form an asteroid one-third of a mile across. More than a century since its completion, work on Culebra Cut has never entirely ceased. With silt constantly accumulating, and frequent small landslides, each day dredging rigs with suction pumps and shovels work up one side of the canal as ships come down the other.

In the green mountains 20 miles northeast of the Culebra Cut, two Panama Canal hydrologists, Modesto Echevers and Johnny Cuevas, stand on a concrete abutment above Lake Alajuela, created by yet another dam, one that had to be built upriver on the Chagres in 1935. The Chagres watershed is one of the rainiest places on Earth, and during the Canal's first two decades, several floods slammed into it. Boat traffic halted for hours while floodgates were opened, lest the pounding of the river cave in its banks. The flood of 1923, which carried entire uprooted mahogany trunks, created a surge on Lake Gatún powerful enough to tip over ships.

Madden Dam, the wall of concrete that holds back the river to form Lake Alajuela, also sends electricity and drinking water to Panama City. But to keep its reservoir from leaking out the sides, engineers had to fill 14 dips in the terrain with earth to create its rim. Down below, massive Lake Gatún is also surrounded by earthen saddle dams. Some are so overgrown with rain forest that an untrained eye can't see that they are artificial—which is why Echevers and Cuevas must come up here every day: to try to stay ahead of nature.

"Everything grows so fast," explains Echevers, a burly man in a blue rain jacket. "When I started doing this, I came here looking for Dam Number 10, and I couldn't find it. Nature had eaten it."

Cuevas nods, eyes closed, recalling many battles with roots that can tear an earthen dam apart. The other enemy is the trapped water itself. During a rainstorm, these men are often here all night, fighting to maintain a balance between holding the Chagres at bay and releasing enough

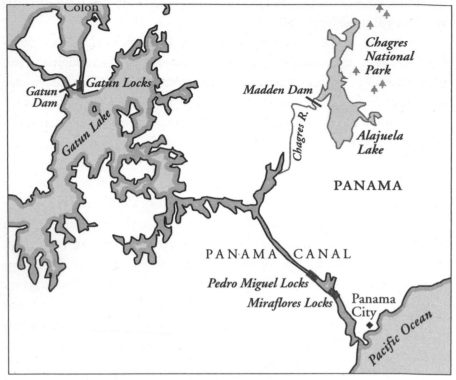

Map of Panama Canal.

MAP BY VIRGINIA NOREY

water through the concrete wall's four floodgates to assure that nothing bursts.

But if one day there were no people around to do that?

Echevers shudders at the thought, because he's seen how the Chagres reacts to rain: "Like a zoo animal that has never accepted its cage. The water loses control. If it was allowed to rise, it would top the dam."

He stops to watch a pickup truck roll across the raised roadway that runs along the top of the dam. "If no one were here to open the flood-gates, the lake would fill up with branches, tree trunks, and garbage, and at some point all that stuff would hit the dam and take the road with it."

Cuevas, his quiet colleague, has been mentally calculating. "The head of the river would be huge when it goes over the top. Like a waterfall, it would erode away the river bottom in front of the dam. One really big flood could collapse the dam."

Even if that never happened, they agree, eventually the spillway gates would rust away. "At that point," says Echevers, "a 20-foot head of water would break free. Drastically."

They look down at the lake where 20 feet below, an eight-foot alligator floats motionless in the dam's shadow, then streaks through the teal blue water as an unlucky terrapin surfaces. Madden Dam's concrete wedge looks too solid to go anywhere. Yet one rainy day, it will likely flop over.

"Even if it survives," says Echevers, "with no one here the Chagres will fill the lake with sediment. At that point, the dam won't matter."

In a chain-link compound where Panama City now spills into the former Canal Zone, Port Captain Bill Huff sits in jeans and a golf shirt before a wall of maps and monitors, guiding evening traffic through the Canal. A U.S. citizen born and raised here—his grandfather, a Canal Zone shipping agent, arrived in the 1920s—he moved to Florida after sovereignty over the Canal passed from the United States to Panama as the clock ticked off the first second of the new millennium. But his 30 years of experience were still in demand, and, now in the employ of Panama, he returns every few months to take a shift.

He switches a screen to a view of Lake Gatún's dam, a low mound of earth 100 feet wide. Its submerged base is 20 times thicker. To the casual observer, there's not much to see. But someone has to be looking all the time.

"There are springs underneath the dam. A couple of small ones have pierced through. If water runs clear, no problem. Clear water means it's coming up through the bedrock." Huff pushes back in his chair and rubs the dark beard circling his chin. "But if water starts bringing dirt with it, then the dam is doomed. In just a matter of hours."

It's hard to imagine. Gatún Dam has a 1,200-foot-thick, theoretically impermeable central core of rock and gravel cemented with liquid clays that are known as fines, sluiced up from the dredged channel below and tamped between two buried rock walls.

"The fines hold the gravel and everything together. They're what starts coming out first. Then the gravel follows, and the dam loses its adhesion."

He opens a long drawer in an old pine desk and pulls out a map tube. Unrolling a yellowed, laminated chart of the isthmus, he points to Gatún

Dam, just six miles from the Caribbean. On the ground, it's an impressive mile-and-a-half long, but on the map it's clearly just a narrow gap compared to the tremendous expanse of water dammed behind it.

The hydrologists Cuevas and Echevers are right, he says. "If not during the first rainy season, within just a few years it would be the end of Madden Dam. *That* lake would all come pouring down into Gatún Lake."

Gatún Lake would then start spilling over the locks on both sides, toward the Atlantic and the Pacific. For a while a casual observer might not notice much, "except maybe unkept grass." The Canal's prim landscaping, still maintained to American military standards, would start to turn lush. But before any palms or figs moved in, a flood would take over.

"Big surges of water would sluice around the locks and scour bypasses into the dirt. Once one of the lock walls started to tumble, that would be the end. All of Gatún Lake could spill." He pauses. "That is, if it hadn't already emptied into the Caribbean. After 20 years with no maintenance, I don't see earthen dams left. Especially Gatún."

At that point, the liberated Chagres River, which drove many French and American engineers crazy and thousands of laborers to their death, would seek its old channel to the sea. With the dams gone, the lakes empty, and the river again headed east, the Pacific side of the Panama Canal would dry up, and the Americas would be reunited.

The last time that happened, 3 million years ago, one of the greatest biological interchanges in Earth's history commenced as North and South American land species began to travel the Central American isthmus, which now joined them.

Until then, the two landmasses had been separated since the supercontinent of Pangea began to break up about 200 million years earlier. During that time, the two separate Americas had embarked on enormously different evolutionary experiments. Like Australia, South America developed a menagerie of marsupial mammals, ranging from sloths to even a lion that carried its young in a pouch. In North America, a more efficient, ultimately triumphant placental path emerged.

This most recent man-made separation has existed for little more than a century—not enough time for any meaningful species evolution, and a canal barely wide enough for two ships to pass each other has hardly been much of a barrier. Still, speculates Bill Huff, until roots work their way into the cracks in the huge, empty concrete boxes that once held ocean-going vessels and finally shatter them, for a few centuries they will be rain-catch holes prowled by panthers and jaguars, as regenerating tapir, white-tailed deer, and anteaters come to drink.

Even longer than those boxes, for a while a big man-made, V-shaped gouge would remain, marking the place where humans undertook, in the words of Theodore Roosevelt after he went to Panama in 1906 to see for himself, "the greatest engineering feat of the ages. The effect of their work," he added, "will be felt while our civilization lasts."

If we disappeared, the words of this larger-than-life American president, who founded a national park system and institutionalized North American imperialism, would prove prophetic. Yet long after the walls of the Culebra Cut cave in, one last larger-than-life monument to Roosevelt's grand vision for the Americas will remain.

❧

IN 1923, SCULPTOR Gutzon Borglum was commissioned to immortalize the greatest American presidents in portraits every bit as imposing as that long-vanished wonder, the Colossus of Rhodes. His canvas was an entire South Dakota mountainside. Along with George Washington, father of the country; Thomas Jefferson, drafter of its Declaration of Independence and Bill of Rights; and Abraham Lincoln, the emancipator and reuniter, Borglum insisted on portraying Theodore Roosevelt, who joined the seas.

The site he selected for what qualifies as the United States' national magnum opus, Mount Rushmore, is a 5,725-foot uplift composed of fine-grained Precambrian granite. When Borglum died in 1941 of a brain hemorrhage, he'd barely begun work on the presidential torsos. But the faces were all indelibly carved in stone; he lived to see the visage of his personal hero, Teddy Roosevelt, officially dedicated in 1939.

He'd even rendered Roosevelt's trademark pince-nez in rock—a rock formed 1.5 billion years ago, among the most resistant on the continent. According to geologists, Mount Rushmore's granite erodes only one inch every 10,000 years. At that rate, barring asteroid collision or a particularly violent earthquake in this seismically stable center of the continent, at least vestiges of Roosevelt's 60-foot likeness, memorializing his Canal, will be around for the next 7.2 million years.

In less time than that, *Pan prior* became us. Should some equally ingenious, confounding, lyrical, and conflicted species appear on Earth again in our aftermath, they may still find T.R's fierce, shrewd gaze fixed intently upon them.

CHAPTER 13

The World Without War

W AR CAN DAMN Earthly ecosystems to hell: witness Vietnam's poisoned jungles. Yet without chemical additives, war curiously has often been nature's salvation. During Nicaragua's Contra War of the 1980s, with shellfish and timber exploitation paralyzed along the Miskito Coast, exhausted lobster beds and stands of Caribbean pine impressively rebounded.

That took less than a decade. And in just 50 years without humans. . . .

THE HILLSIDE IS heavily booby-trapped, which is why Ma Yong-Un admires it. Or rather, he admires the mature stands of daimyo oak, Korean willow, and bird cherry growing wherever land mines have kept people out.

Ma Yong-Un, who coordinates international campaigns at the Korean Federation for Environmental Movement, is climbing through cottony November fog in a white propane-powered Kia van. His companions are conservation specialist Ahn Chang-Hee, wetlands ecologist Kim Kyung-Won, and wildlife photographers Park Jong-Hak and Jin Ik-Tae. They've just cleared a South Korean military checkpoint, slaloming through a maze of black and yellow concrete barriers as they entered this restricted area. The guards, in winter camouflage fatigues, set aside their M16s to greet the KFEM team—since the last time they were here, a year earlier, a sign was added stating that this post is also an environmental checkpoint for preservation of red-crowned cranes.

While waiting for their paperwork, Kim Kyung-Won had made note of several gray-headed woodpeckers, a pair of long-tailed tits, and the bell-like singing of a Chinese bulbul in the dense brush around the checkpoint. Now, as the van ascends, they flush a brace of ring-necked pheasants and several azure-winged magpies, beautiful birds no longer common elsewhere in Korea.

They have entered a strip of land five kilometers deep that lies just below South Korea's northern limit, called the Civilian Control Zone. Nearly no one has lived in the CCZ for half a century, although farmers have been permitted to grow rice and ginseng here. Five more kilometers of dirt road, flanked by barbed wire filled with perching turtle doves and hung with red triangles warning of more minefields, and they reach a sign in Korean and English that says they are entering the Demilitarized Zone.

The DMZ, as it is called even in Korea, is 151 miles long and 2.5 miles wide, and has been a world essentially without people since September 6, 1953. A final exchange of prisoners had ended the Korean War—except, like the conflict that tore Cyprus in two, it never really ended. The division of the Korean Peninsula had begun when the Soviet Union declared war on Japan late in World War II, on the same day that the United States dropped a nuclear warhead on Hiroshima. Within a week, that war was over. An agreement by the Americans and the Soviets to split the administration of Korea, which Japan had occupied since 1910, became the hottest point of contact for what became known as the Cold War.

Abetted by its Chinese and Soviet communist mentors, North Korea invaded the South in 1950. Eventually, United Nations forces pushed them back. A 1953 truce ended what had become a stalemate along the original dividing line, the 38th parallel. A strip two kilometers on either side of it became the no-man's-land known as the Demilitarized Zone.

Much of the DMZ runs through mountains. Where it follows the courses of rivers and streams, the actual demarcation line is in bottomland where, for 5,000 years before the hostilities began, people grew rice. Their abandoned paddies are now sown thickly with land mines. Since the armistice in 1953, other than brief military patrols or desperate, fleeing North Koreans, humans have barely set foot here.

In their absence, the netherworld between these enemy doppelgängers has filled with creatures that had practically nowhere else to go. One of the world's most dangerous places became one of its most important—though inadvertent—refuges for wildlife that might otherwise have disappeared. Asiatic black bears, Eurasian lynx, musk deer, Chinese water deer, yellow-throated marten, an endangered mountain goat known as the goral, and the nearly vanished Amur leopard cling here to what may only be temporary life support—a slender fraction of the necessary range for a genetically healthy population of their kind. If everything north and south of Korea's DMZ were suddenly to become a world without humans as well, they might have a chance to spread, multiply, reclaim their former realm, and flourish.

Ma Yong-Un and his conservation companions have no recollection of Korea without this geographic paradox binding its midriff. Now in their thirties, they were born in a nation that grew from poverty to prosperity while they themselves were growing. Immense economic success has made millions of South Koreans believe—like Americans, Western Europeans, and Japanese before them—that they can have everything. For these young men, that means having their country's wildlife, too.

They arrive at a fortified observation bunker where South Korea has cheated. Here, the 151 miles of double fencing topped with coiled razor wire makes a sharp northward jog, following a promontory nearly one kilometer before looping back. That's nearly half the distance that the truce obliges the two Koreas to maintain from the Demarcation Line, a faint string of posts down the DMZ's middle that neither side is ever to approach.

"They do it, too," Ma Yong-Un explains. Any place where a landform offers a view too irresistible to pass up, both sides seem to welcome opportunities to encroach and stare the other down. The camouflage paint on this artillery placement's cinder blocks serves not to conceal but to display, like a belligerent cock bristling with threats and munitions in lieu of combs and feathers.

At the promontory's northern edge, the DMZ opens into rugged fullness and vast emptiness for miles in either direction. Although each side has held fire since 1953, large loudspeakers atop South Korea's positions have blasted regular insults, military anthems, and even strident themes

Korean DMZ.
PHOTO BY ALAN WEISMAN.

like the *William Tell Overture* across the divide. The din has bounced off North Korean mountainsides that, over the decades, have been increasingly stripped bare for firewood. The inevitable tragic erosion has led to flooding, agricultural disasters, and famine. Should this entire peninsula one day be bereft of people, its ravaged northern half will take far longer to resuscitate biologically, while its southern half will leave far more infrastructure for nature to disassemble.

Below, in the buffer separating these vast extremes, are 5,000-year-old rice paddies that have reverted to wetlands during the last half-century. As the Korean naturalists watch, cameras and spotting scopes poised, over the bulrushes glides a dazzling white squadron, 11 fliers in perfect formation.

And in perfect silence. These are living Korean national icons: red-crowned cranes—the largest, and, next to whooping cranes, rarest on Earth. They're accompanied by four smaller white-naped cranes, also endangered. Just in from China and Siberia, the DMZ is where most of them winter. If it didn't exist, they probably wouldn't either.

They touch down lightly, disturbing no buried hair-triggers. Revered in Asia as sacred portents of luck and peace, the red-crowned cranes are blissfully oblivious trespassers who've wandered into the incandescent tension of 2 million troops faced off across this accidental wildlife sanctuary in bunkers every few dozen meters, mortars poised.

"Babies," Kyung-Won whispers, and the lenses fix on two juvenile cranes wading in a streambed, their long bills rooting underwater for tubers, their crowns still juvenile brown. Only around 1,500 of these birds still exist, and each new birth is momentous.

Behind them, in a North Korean version of the Hollywood sign, the hills sprout whitewashed Korean characters that proclaim the supremacy of Dear Leader Kim Jong-il and loathing for America. Their enemies retort with giant marquees whose thousands of lightbulbs flash messages visible for miles about the good life in the capitalist South. Every few hundred meters between observation posts bursting with propaganda is another armed bunker, with eyes peering through a slit at some opposite number across the chasm. The confrontation has burned through three generations of enemies now, many of them blood relatives.

Through all this menace float the cranes, landing in the sunny flats on both sides of the demarcation line to serenely graze on reeds. None of these men, rapt at the sight of such magnificent winged eminences, would ever admit to praying against peace, but the truth is that if not for the seething hostilities that keep this zone clear, these birds would likely face extinction. Just to the east, the suburbs of Seoul—a juggernaut approaching 20 million *Homo sapiens*—roll ever northward, banging into the CCZ, with developers poised to invade this tantalizing real estate whenever the concertina wire comes down. And North Korea, edging toward China's example, has collaborated with its capitalistic archfoes on a border industrial megapark to tap its most abundant resource: hungry multitudes who will work cheap—and who will need housing.

The ecologists spend an hour watching the regal, nearly five-foot-tall birds in their element. All the while, they themselves are under the unblinking scrutiny of cheerless soldiers charged to defend the border. One approaches to inspect their tripod-mounted, 40-power Swarovski spotting scope. They show him the cranes. As he squints, with his muzzle-mounted grenade launcher pointing skyward, tinted afternoon shadows slant across North Korea's bare mountainsides. A shaft of sunlight spears a white, battle-scarred ridge called T-Bone Hill, which juts up from the contested plain between the two halves of Korea. The soldier tells them how many heroes died defending it, and how many more of the hated enemy were slain.

They've heard this before. "Besides the difference between North and South Korea, you should tell people about the ecosystem we share," Ma Yong-Un replies. He points to a water buck ascending the grassy slope. "One day this will all be one country, but there will still be reason to protect it."

They return through a long, flat Civilian Control Zone valley carpeted with rice stubble. The soil is scored into herringbone furrows separated by glinting mirrors of early snowmelt that will re-freeze by nightfall. By December, temperatures will descend to -20°F The sky is hatched with patterns echoing the ploughed geometrics below as lines of cranes soar in, joined by great airborne wedges of thousands of geese.

As the birds descend for an afternoon meal of rice harvest remains, the group stops for photographs and a quick census. There are 35 red-crowned cranes, looking straight out of a Japanese silk painting: glowing white, with cherry skullcaps and black necks. There are also 95 pink-legged, white-naped cranes. There are three species of geese: upland, bean, and some rare spotted snow geese, all protected from hunting in South Korea, so many that no one bothers counting them.

Thrilling as it was to spot cranes in the recovering natural DMZ wetlands, it's far easier in these adjacent tilled lands, where they can feast on grains missed by mechanized harvesters. Would these birds benefit or suffer if humans were to disappear? Red-crowned cranes evolved to nibble reed shoots, but by now thousands of generations have been fed in human-engineered wetlands called rice paddies. If there were no more farmers, and if the bountiful rice fields of the CCZ also revert to marshes, would crane and geese populations decline?

"A rice paddy is not an ideal ecosystem for these cranes," declares Kyung-Won, looking up from his spotting scope. "They need roots, not just grain. So many wetlands have turned into farms, they have no choice but to eat this for energy to survive winter."

In the DMZ's abandoned rice paddies, not enough reeds and canary grass have reappeared to support even these critically diminished populations, because both Koreas have built dams upstream. "Even in winter they pump water to grow vegetables in greenhouses, when the aquifers should be replenishing with snowfall," says Kyung-Won.

If there were no agriculture trying to feed 20 million humans in Seoul, let alone North Korea, pumps that defy the very seasons would be stilled. Water would return, and wildlife with it. "For plants and animals, it would be such a relief," says Kyung-Won. "A paradise."

Like the DMZ itself: a killing place that became a haven to nearly vanished Asian creatures. Even the all-but-extinct Siberian tiger is rumored to hide here, though that may just be wishful dreaming. What these young naturalists want is exactly what their counterparts in Poland and Belarus beg for: a peace park, transmuted from a war zone. A coalition of international scientists called the DMZ Forum has tried to convince politicians of the potential for a face-saving peace, and even profit, if Korea's two enemies together consecrate the one good thing they share.

"Think of a Korean Gettysburg and Yosemite rolled together," says DMZ Forum co-founder, Harvard biologist E. O. Wilson. Even with the expensive prospect of clearing all the land mines, Wilson believes that tourism revenue could trump agriculture or development. "A hundred years from now, of all the things that happened here in the last century, what will matter most will be that park. It will be the legacy most treasured by the Korean people, and an example for the rest of the world to follow."

It's a sweet vision, but one on the verge of being swallowed by subdivisions that already crowd the DMZ. The Sunday after he returns to Seoul, Ma Yong-Un visits the Hwa Gye Sah Temple in the mountains north of town, one of Korea's oldest Buddhist sanctuaries. In a pavilion adorned with carved dragons and gilded Bodhisattvas, he hears disciples chant the Diamond Sutra, in which Buddha teaches that all is like a dream, an illusion, a bubble, a shadow. Like the dew.

"The world is impermanent," the gray-robed head monk, Hyon Gak Sunim, tells him afterward. "Like our body, we must let go of it." Yet, he assures Ma Yong-Un, to try to preserve the planet isn't a Zen paradox. "The body is essential for enlightenment. We have an obligation to take care of ours."

But the sheer number of human bodies now makes caring for the Earth a particularly perplexing koan. Even the once-sacrosanct tranquility of Korea's temples is under assault. To shorten the commute into Seoul from outlying suburbs, an eight-lane tunnel is being dug directly beneath this one.

"In this century," insists E. O. Wilson, "we'll develop an ethic of letting population gradually subside, until we reach a world with far less human impact." He says this with the conviction of a scientist so steeped in probing the resilience of life that he claims it for his own species as well. But if land mines can be swept for tourists, real estate mongers will scheme for the same prime property. If a compromise results in developments surrounding a token history-nature theme park, the only viable species left in the DMZ will likely be our own.

Until, that is, the two Koreas—together nearly 100 million humans on a peninsula the size of Utah—finally topple under the weight of their resident *Homo* population. If, however, people simply vanish first, even though the DMZ may be too marginal to sustain Siberian tigers, "a few," muses Wilson, "still prowl the mountains of the North Korean–Chinese borderlands." His voice warms as he envisions them multiplying and fanning across Asia while lions work their way up southern Europe.

"Pretty quickly there would be a tremendous spread of remaining megafauna," he continues. "Especially the carnivores. They'd make short work of our livestock. After a couple hundred years, few domestic animals would remain. Dogs would go feral, but they wouldn't last long: they'd never be able to compete. There would be a huge shakeout involving species introduced wherever there's been human disturbance."

In fact, bets E. O. Wilson, all human attempts to improve on nature, such as our painstakingly bred horses, would revert to their origins. "If horses even survived, they would devolve back to Przewalski's horse"—the only true wild horse remaining, of the Mongolian steppes.

"The plants, crops, and animal species man has wrought by his own hand would be wiped out in a century or two. Many others would also be gone, but there would still be birds and mammals. They'd just be smaller. The world would mostly look as it did before humanity came along. Like a wilderness."

CHAPTER 14

Wings Without Us

1. Food

AT THE WESTERN end of the Korean DMZ, on a mud-pancake island in the Han River estuary, nests one of the rarest large birds of all: the black-faced spoonbill. Only 1,000 remain on Earth. North Korean ornithologists have clandestinely warned colleagues across the river that their hungry comrade citizens swim out to poach spoonbill eggs. The South Korean hunting ban is no help, either, to geese that land north of the DMZ. Nor do cranes there banquet on rice kernels spilled by mechanized harvesters. The reaping in North Korea is all by hand, and people take even the smallest grains. Nothing is left for birds.

In a world without humans, what will be left for birds? What will be left *of* birds? Of the more than 10,000 species that have coexisted with us, ranging from hummingbirds that weigh less than a penny to 600-pound wingless moas, about 130 have disappeared. That is barely more than 1 percent, almost an encouraging figure if some of these losses hadn't been so sensational. Moas stood 10 feet tall and weighed twice as much as an African ostrich. They were extinguished within two centuries by Polynesians who sometime around AD 1300 colonized the last major planetary landmass that humans discovered, New Zealand. By the time Europeans appeared some 350 years later, piles of big bird bones and Maori legends were all that remained.

Other massacred, flightless birds include the dodo of the Indian Ocean's Mauritius Island, famously clubbed and cooked to death within a hundred years by Portuguese sailors and Dutch settlers it never learned to fear. Because the penguin-like great auk's range stretched across the upper Northern Hemisphere, it took longer, but hunters from Scandinavia to Canada managed to exterminate them anyhow. The moa-nalo—flightless, oversized ducks that ate leaves—went extinct long ago in Hawaii; we know little about them, other than who killed them.

The most stunning avicide of all, just a century ago, is still hard to fathom in its enormity. Like hearing astronomers explain the entire universe, its lesson gets lost because its subject, when it was alive, literally exceeded our horizons. The postmortem of the American passenger pigeon is so fertile with portents that just a brief glance warns—screams, in fact—that anything we consider limitless probably isn't.

Long before we had poultry factories to mass-produce chicken breasts by the billion, nature did much the same for us in the form of the North American passenger pigeon. By anyone's estimate, it was the most abundant bird on Earth. Its flocks, 300 miles long and numbering in the billions, spanned horizons fore and aft, actually darkening the sky. Hours could go by, and it was as though they hadn't passed at all, because they kept coming. Larger, far more striking than the ignoble pigeons that soil our sidewalks and statuary, these were dusky blue, rose breasted, and apparently delicious.

They ate unimaginable quantities of acorns, beechnuts, and berries. One of the ways we slew them was by cutting their food supply, as we sheared forests from the eastern plains of the United States to plant our own food. The other was with shotguns, spraying lead pellets that could down dozens with a single blast. After 1850, with most of the heartland forest gone to farms, hunting passenger pigeons was even easier, as millions of them roosted together in the remaining trees. Boxcars stuffed with them arrived daily in New York and Boston. When it finally became apparent that their unthinkable numbers were actually dropping, a kind of madness drove hunters to slaughter them even faster while they were still there to kill. By 1900, it was over. A miserable few remained caged in a Cincinnati zoo, and by the time zookeepers realized what they had, nothing could be done. The last one died before their eyes in 1914.

In succeeding years, the parable of the passenger pigeon was retold often, but its moral could only be heeded in part. A conservation movement

Passenger pigeon. *Ectopistes migratorius.*
ILLUSTRATION BY PHYLLIS SAROFF

founded by hunters themselves, Ducks Unlimited, has bought millions of acres of marshland to insure that no game species they value will be without places to land and breed. However, in a century in which humans proved more inventive than during the rest of *Homo sapiens* history combined, protecting life on the wing became more complicated than simply making game-bird hunting sustainable.

2. Power

The Lapland longspur isn't commonly known to North Americans, because its behavior isn't quite what we expect from migratory birds. Its summer and breeding grounds are in the high Arctic, so just as more familiar songbirds head toward the equator and beyond, Lapland longspurs arrive to spend the winter in the great plains of Canada and the United States.

They're pretty little black-faced, finch-sized birds with white half masks and russet patches on their wings and nape, but we mostly see them at a distance: hundreds of indistinct, small birds swirling in the winter prairie wind, picking over fields. On the morning of January 23, 1998, however, they were easy to see in Syracuse, Kansas, because nearly 10,000 were lying frozen on the ground. During a storm the previous evening, a flock crashed into a cluster of radio-transmission towers. In the fog and blowing snow, the only things visible were red, blinking lights, and the longspurs apparently headed for them.

Neither the circumstances nor the numbers of their deaths were particularly unusual, although the toll for a single evening was possibly high. Reports of dead birds heaped around the bases of TV antennae started getting ornithologists' attention in the 1950s. By the 1980s, estimates of 2,500 deaths per tower, per year, were appearing.

In 2000, the U.S. Fish and Wildlife Service reported that 77,000 towers were higher than 199 feet, which meant that they were required to have warning lights for aircraft. If calculations were correct, that meant that nearly 200 million birds collided fatally with towers each year in the United States alone. In fact, those figures had already been usurped, because cell phone towers were being erected so fast. By 2005, there were 175,000 of those. Their addition would raise the annual toll to half a billion dead birds—except that this number was still based on scant data and on guesses, because scavengers get to most feathered victims before they're found.

From ornithology labs east and west of the Mississippi, graduate students were sent on grisly night missions to transmitter towers to recover the carcasses of red-eyed vireos, Tennessee warblers, Connecticut warblers, orange-crowned warblers, black-and-white warblers, ovenbirds, wood thrushes, yellow-billed cuckoos . . . the lists became an increasingly thorough compendium of North American birds, including rare species like the red-cockaded woodpecker. Especially prominent were birds that migrate, and especially those that travel at night.

One is the bobolink, a black-breasted, buff-backed plains songbird that winters in Argentina. By studying its eyes and brains, bird physiologist Robert Beason has detected evolutionary traits that unfortunately turned lethal in the age of electronic communications. Bobolinks and other migrants carry built-in compasses—particles of magnetite in their heads, with which they orient to the Earth's magnetic field. The mechanism to switch them on involves their optics. The short end of the spectrum—purples, blues, and greens—apparently triggers their navigational cues. If only longer red waves are present, they grow disoriented.

Beason's observations also suggest that migrating birds evolved to fly toward light in foul weather. Until electricity, this meant the moon, which would put them out of harmful weather's way. Thus, a pulsating tower bathed in a red glow whenever fog or blizzard blots out everything else is as seductive and deadly to them as wailing Sirens to Greek sailors. With their homing magnets befuddled by a transmitter's electromagnetic fields, they end up circling its towers, whose guy wires become the blades of a giant bird blender.

In a world without humans, the red lights will blink off as broadcasts cease; a billion daily cellular conversations will disconnect, and several billion more birds will be alive a year later. But as long as we're still here, transmission towers are only the beginning of the unintended carnage human civilization perpetrates on feathered creatures we don't even eat.

A different kind of tower—frameworks of steel lattice averaging 150 feet tall, spaced every 1,000 feet or so—marches the length and breadth and diagonally across every continent save Antarctica. Suspended between these structures are aluminum-clad high-tension cables bearing millions of sizzling volts from power plants to our energy grids. Some are three inches thick; to save weight and cost, all are uninsulated.

There's enough wire in North America's grid alone to reach the moon and back, and nearly back again. With the clearing of forests, birds learned to perch on telephone and power lines. As long as they don't complete a circuit with another wire or with the ground, they don't electrocute themselves. Unfortunately, the wings of hawks, eagles, herons, flamingos, and cranes can span two wires at once, or brush an uninsulated transformer. The result is no mere shock. A raptor's beak or feet can melt right off, or its feathers can ignite. Several captive-bred California condors have died exactly this way on being released, as have thousands of bald and golden eagles. Studies in Chihuahua, Mexico, show that new steel power poles act like giant ground wires, so that even smaller birds end up on the piles of dead hawks and turkey vultures below.

Other research suggests that more birds die by simply colliding with power lines than from being zapped by them. But even without webs of live wires, the most serious traps for migratory birds await in tropical America and Africa. So much land there has been cleared for agriculture, much of it for export, that each year there are fewer roosting trees to ease the journey, and fewer safe wetlands where waterfowl can pause. As with climatic change, the impact is hard to quantify, but in North America and Europe, the numbers of some songbird species have fallen by two-thirds since 1975.

Without humans, some semblance of those wayside forests will return within a few decades. Two other major perpetrators of songbird loss—acid rain, and insecticide use on corn, cotton, and fruit trees—will end immediately when we're gone. The resurgence of bald eagles in North America after DDT was banned bodes hopeful for creatures that cope with residual traces of our better life through chemistry. However, while DDT is toxic at a few parts per million, dioxins become dangerous at just 90 parts per *trillion*—and dioxins may remain until the end of life itself.

In separate studies, two U.S federal agencies estimate that 60 to 80 million birds also annually end up in radiator grilles or as smears on windshields of vehicles racing down highways that, just a century ago, were slow wagon trails. High-speed traffic would end when we do, of course. However, the worst of all man-made menaces to avian life is totally immobile.

Well before our architecture tumbles, its windows will mostly be gone, and one reason will be repeated pounding from inadvertent avian kamikazes. While Muhlenberg College ornithologist Daniel Klem was earning his doctorate, he enlisted suburban New York and southern Illinois residents to record the numbers and kinds of birds crashing into that post–World War II home builder's icon, the plate glass picture window.

"Windows are not recognized as obstacles by birds," Klem tersely notes. Even when he stood them in the middle of fields, free of surrounding walls, birds failed to notice them until the final, violent second of their lives.

Big birds, little birds, old or young, male or female, day or night—it didn't matter, Klem discovered over two decades. Nor did birds discriminate between clear glass and reflective panes. That was bad news, given the late-20th-century spread of mirrored high-rises beyond city centers, out to exurbs that migrating birds recall as open fields and forests. Even nature park visitor centers, he says, are often "literally covered with glass, and these buildings regularly kill birds that the public comes to see."

Klem's 1990 estimate was 100 million annual bird necks broken from flying into glass. He now believes that 10 times that many—1 billion in the United States alone—is probably too conservative. There are about 20 billion total birds in North America. With another 120 million taken each year by hunting—that same pastime that snuffed mammoths and passenger pigeons—these numbers begin to add up. And there is still one more scourge that man has wreaked on birdlife, one that will outlive us—unless it runs out of birds to devour.

3. The Pampered Predator

Wisconsin wildlife biologists Stanley Temple and John Coleman never needed to leave their home state to draw global conclusions from their field research during the early 1990s. Their subject was an open secret—a topic hushed because few will admit that about one-third of all households, nearly everywhere, harbor one or more serial killers. The villain is the purring mascot that lolled regally in Egyptian temples and does the same on our furniture, accepting our affection only when it pleases, exuding inscrutable calm whether awake or asleep (as it spends more than half its life), beguiling us to see to its care and feeding.

Once outside, however, *Felis silvestris catus* drops its subspecies surname and starts stalking as it reverts to being *F. silvestris*—wild cat—genetically identical to small native wildcats still found, though seldom seen, in Europe, Africa, and parts of Asia. Although cunningly adapted over a few thousand years to human comforts—cats that never venture outdoors generally live far longer—domestic cats, Temple and Coleman report, never lost their hunting instincts.

Possibly, they sharpened them. When European colonists first brought them, American birds had never before seen this sort of silent, tree-scaling, pouncing predator. America has bobcats and Canadian lynx, but this fecund invasive feline species was a quarter-size version—a frightening, perfect fit for the enormous population of songbirds. Like Clovis Blitzkriegers, cats killed not only for sustenance, but also seemingly for the sheer pleasure of it. "Even when fed regularly by people," Temple and Coleman wrote, "a cat continues hunting."

In the past half-century, as the world's human population doubled, the number of cats did so much faster. In U.S. Census Bureau pet figures,

Temple and Coleman found that from merely 1970 to 1990, America's cat count rose from 30 to 60 million. The actual total, however, must also include feral cats that form urban colonies and rule barnyards and woodlands in far greater densities than comparable-sized predators like weasels, raccoons, skunks, and foxes, which have no access to protective human shelter.

Various studies credit alley cats with up to 28 kills per year. Farm cats, Temple and Coleman observed, get many more than that. Comparing their findings with all the available data, they estimated that in rural Wisconsin, around 2 million free-ranging cats kill at minimum 7.8 million, but probably upwards of 219 million, birds per year.

That's in rural Wisconsin alone.

Nationwide, the number likely approaches the billions. Whatever the actual sum may be, cats will do very well in a world without the people who took them to all the continents and islands they didn't already inhabit, where they now outnumber and out-compete other predators their own size. Long after we're gone, songbirds must deal with the progeny of these opportunists that trained us to feed and harbor them, disdaining our hapless appeals to come when we call, bestowing just enough attention so we feed them again.

❦

IN FOUR DECADES of birding, ornithologist Steve Hilty, author of two of the world's thickest field guides (to the birds of Colombia and Venezuela), has seen some strange human-caused changes. He's watching one of them from the shore of a glacial lake just outside the town of Calafate in southern Argentina, near the Chilean border: kelp gulls from Argentina's Atlantic coast, which have now spread across the country and grown 10 times more abundant simply from scavenging landfills. "I've watched them follow human trash across Patagonia like house sparrows going after spilled grain. Now there are far fewer geese on lakes, because gulls prey on them."

In a world without people's garbage, guns, and glass, Hilty predicts a reshuffling of populations back toward their former balance. Some things may take longer, as temperature shifts have done funny things to their ranges. Some brown thrashers in the southeastern United States today don't bother to migrate, and red-wing blackbirds have even passed up Central America to winter over in southern Canada, where they now encounter a classic southern U.S. species, the mockingbird.

As a professional birding guide, Hilty has watched the decline of songbirds tilt into a plunge that has even nonbirders noticing the deepening silence. Among the missing in his native Missouri is our only blue-backed, white-throated warbler. Cerulean warblers used to depart the Ozarks each fall for mid-elevation Andean forests in Venezuela, Colombia, and Ecuador. With more of those being cut each year for coffee—or coca—hundreds of thousands of arriving birds must funnel into an ever-shrinking wintering ground, where there isn't enough to feed them all.

One thing still heartens him: "In South America, very few birds have actually gone extinct." That is huge, because South America has more bird species than anywhere. When the Americas were joined 3 million years ago, just below the juncture at Panama was mountainous Colombia, poised to be a giant species trap, with every niche from coastal jungle to alpine moor. Colombia's number-one rank—more than 1,700 bird species—is sometimes challenged by ornithologists in Ecuador and Peru, which means that even more vital habitat still remains. Yet too often, just barely so: Ecuador's white-winged brush finch now lives in only one Andean valley. Northeast Venezuela's gray-headed warbler is confined to a single mountaintop. Brazil's cherry-throated tanager is found on just a single ranch north of Rio.

In a world without people, the birds that survived would soon reseed South American trees that were displaced by rows of that Ethiopian immigrant *Coffea arabica*. With no one there to weed, new seedlings would battle coffee bushes for nutrients. In a few decades, shade from their canopies would slow the interloper's growth, and their roots would strangle it until it choked.

Coca plants—native to the highlands of Peru and Bolivia, but needing chemical help anywhere else—won't last two seasons in Colombia without men to tend them. But dead coca fields, like cattle pastures, will leave a checkerboard of empty bald spots where forest clandestinely came down. One of Hilty's biggest concerns is for small Amazonian birds so adapted to dense cover that they can't tolerate bright light. Many fail because they won't cross open areas.

A scientist named Edwin Wills discovered that, just after the Panama

Canal was completed. As Lake Gatún filled, some mountains ended up as islands. The biggest, 3,000-acre Barro Colorado, became a research laboratory for the Smithsonian's Tropical Studies Institute. Wills began to study foraging antbirds and ground cuckoos—until suddenly they were gone.

"Three thousand acres weren't enough to sustain a population of species that won't cross open water," says Steve Hilty. "In forest islands separated by pastures, it's the same."

The birds that manage to survive on islands, as Charles Darwin momentously observed among finches in the Galápagos, can adapt so tightly to local conditions that they become species unto themselves, found nowhere else. Those conditions explode, however, once humans arrive with their pigs, goats, dogs, cats, and rats.

In Hawaii, all the roast feral pig devoured in luaus can't keep up with the mayhem their rooting wreaks on forests and bogs. To protect exotic sugarcane from being eaten by exotic rats, in 1883 Hawaiian growers imported the exotic mongoose. Today, rats are still around: the favorite food of both the rat and the mongoose is the eggs of the few native geese and nesting albatrosses left on Hawaii's main islands. In Guam, just after World War II, a U.S. transport plane landed bearing stowaway Australian brown tree snakes in its wheel-wells. Within three decades, along with several native lizards, more than half the island's bird species were extinct, and the rest designated uncommon or rare.

When we humans become extinct ourselves, part of our legacy will live on in the predators we introduced. For most, the only constraints on their rampant proliferation have been the eradication programs with which we've tried to undo our damage. When we go, those efforts go with us, and rodents and mongooses will inherit most of the South Pacific's lovely isles.

Although albatrosses spend most of their lives on their majestic wings, they still must land in order to breed. Whether they will still have enough safe places to do so is uncertain, whether we're gone or not.

CHAPTER 15

Hot Legacy

1. The Stakes

As befits a chain reaction, it happened very fast. In 1938, a physicist named Enrico Fermi went from Fascist Italy to Stockholm to accept the Nobel Prize for his work with neutrons and atomic nuclei—and kept going, defecting with his Jewish wife to the United States.

That same year, word leaked that two German chemists had split uranium atoms by bombarding them with neutrons. Their work confirmed Fermi's own experiments. He had guessed correctly that when neutrons cracked an atomic nucleus, they would set more neutrons free. Each would scatter like a subatomic shotgun pellet, and with enough uranium handy, they would find more nuclei to destroy. The process would cascade, and a lot of energy would be released. He suspected Nazi Germany would be interested in that.

On December 2, 1942, in a squash court beneath the stadium at the University of Chicago, Fermi and his new American colleagues produced a controlled nuclear chain reaction. Their primitive reactor was a beehive-shaped pile of graphite bricks laced with uranium. By inserting rods coated with cadmium, which absorbs neutrons, they could moderate the exponential shattering of uranium atoms to keep it from getting out of hand.

Less than three years later, in the New Mexico desert, they did just the opposite. The nuclear reaction this time was intended to go completely out of control. Immense energy was released, and within a month the act was repeated twice, over two Japanese cities. More than 100,000 people died instantly, and the dying continued long after the initial blast. Ever since, the human race has been simultaneously terrified and fascinated by the double deadliness of nuclear fission: fantastic destruction followed by slow torture.

If we left this world tomorrow—assuming by some means other than blowing ourselves to bits—we would leave behind about 30,000 intact nuclear warheads. The chance of any exploding with us gone is effectively zero. The fissionable material inside a basic uranium bomb is separated into chunks that, to achieve the critical mass necessary for detonation, must be slammed together with a speed and precision that don't occur in nature. Dropping them, striking them, plunging them in water, or rolling a boulder over them would do nothing. In the tiny chance that the polished surfaces of enriched uranium in a deteriorated bomb actually met, unless forced together at gunshot speed, they would fizzle—albeit in a very messy way.

A plutonium weapon contains a single fissionable ball that must be forcibly, exactly compressed to at least twice its density to explode. Otherwise, it's simply a poisonous lump. What *will* happen, however, is that bomb housings will ultimately corrode, exposing the hot innards of these devices to the elements. Since weapons-grade plutonium-239 has a half-life of 24,110 years, even if it took an ICBM cone 5,000 years to disintegrate, most of the 10 to 20 pounds of plutonium it contained would not have degraded. The plutonium would throw off alpha particles—clumps of protons and neutrons heavy enough to be blocked by fur or even thick skin, but disastrous to any creature unlucky enough to inhale them. (In humans, 1 millionth of a gram can cause lung cancer.) In 125,000 years, there would be less than a pound of it, though it would still be plenty lethal. It would take 250,000 years before the levels were lost in the Earth's natural background radiation.

At that point, however, whatever lives on Earth would still have to contend with the still-deadly dregs of 441 nuclear plants.

2. Sunscreen

When big, unstable atoms like uranium decay naturally, or when we rip them apart, they emit charged particles and electromagnetic rays similar to the strongest X-rays. Both are potent enough to alter living cells and DNA. As these deformed cells and genes reproduce and replicate, we sometimes get another kind of chain reaction, called cancer.

Since background radiation is always present, organisms have adjusted accordingly by selecting, evolving, or sometimes just succumbing. Anytime we raise the natural background dosage, we force living tissue to respond. Two decades prior to harnessing nuclear fission, first for bombs, then for power plants, humans had already let one electromagnetic genie loose—the result of a goof we wouldn't recognize until nearly 60 years later. In that instance, we didn't coax radiation out but let it sneak in.

That radiation was ultraviolet, a considerably lower energy wave than the gamma rays emitted from atomic nuclei, but it was suddenly present at levels unseen since the beginning of life on Earth. Those levels are still rising, and although we have hopes to correct that over the next half century, our untimely departure could leave them in an elevated state far longer.

Ultraviolet rays helped to fashion life as we know it—and, oddly enough, they created the ozone layer itself, our shield against too much exposure to them. Back when the primordial goo of the planet's surface was being pelted with unimpeded UV radiation from the sun, at some pivotal instant—perhaps sparked by a jolt of lightning—the first biological mix of molecules jelled. Those living cells mutated rapidly under the high energy of ultraviolet rays, metabolizing inorganic compounds and turning them into new organic ones. Eventually, one of these reacted to the presence of carbon dioxide and sunlight in the primitive atmosphere by giving off a new kind of exhaust: oxygen.

That gave ultraviolet rays a new target. Picking off pairs of oxygen atoms joined together—O_2 molecules—they split them apart. The two singles would immediately latch onto nearby O_2 molecules, forming O_3: ozone. But UV easily breaks the ozone molecule's extra atom off, reforming oxygen; just as quickly, that atom sticks to another pair, forming more ozone until it absorbs more ultraviolet and spins off again.

Gradually, beginning about 10 miles above the surface, a state of equilibrium emerged: ozone was constantly being created, pulled apart, and recombined, and thus constantly occupying UV rays so that they never reached the ground. As the layer of ozone stabilized, so did the life on Earth it was shielding. Eventually, species evolved that could never have tolerated the former levels of UV radiation bombardment. Eventually, one of them was us.

In the 1930s, however, humans started undermining the oxygen-ozone balance, which had remained relatively constant since soon after life began. That's when we started using Freon, the trademark name for chlorofluorocarbons, the man-made chlorine compounds in refrigeration. Called CFCs for short, they seemed so safely inert that we put them into aerosol cans and asthma-medication inhalers, and blew them into polymer foams to make disposable coffee cups and running shoes.

In 1974, University of California–Irvine chemists F. Sherwood Rowland and Mario Molina began to wonder where CFCs went once those refrigerators or materials broke down, since they were so impervious to combining with anything else. Eventually, they decided that hitherto indestructible CFCs must be floating to the stratosphere, where they would finally meet their match in the form of powerful ultraviolet rays. The molecular slaughter would free pure chlorine, a voracious gobbler of loose oxygen atoms, whose presence kept those same ultraviolet rays away from Earth.

No one paid Rowland and Molina much heed until 1985, when Joe Farman, a British researcher in Antarctica, discovered that part of the sky was missing. For decades, we'd been dissolving our UV screen by soaking it with chlorine. Since then, in unprecedented cooperation, the nations of the world have tried to phase out ozone-eating chemicals. The results are encouraging, but still mixed: Ozone destruction has slowed, but a black market in CFCs thrives, and some are still legally·produced for "basic domestic needs" in developing countries. Even the replacements we commonly use today, hydrochlorofluorocarbons, HCFCs, are simply milder ozone-destroyers, scheduled to be phased out themselves—though the question of with what isn't easily answered.

Quite apart from ozone damage, both HCFCs and CFCs—and their most common chlorine-free substitute, hydrofluorocarbons, HFCs—have many times the potential of carbon dioxide to exacerbate global warming. The use of all these alphabetical concoctions will stop, of course, if human activity does, but the damage we did to the sky may last a lot longer. The best current hope is that the South Pole's hole, and the thinning of the ozone layer everywhere else, will heal by 2060, after destructive substances are exhausted. This assumes that something safe will have replaced them, and that we'll have found ways to get rid of existing supplies that haven't yet drifted skyward. Destroying something designed to be indestructible, however, turns out to be expensive, requiring sophisticated, energy-intensive tools such as argon plasma arcs and rotary kilns that aren't readily available in much of the world.

As a result, especially in developing countries, millions of tons of CFCs are still used or linger in aging equipment, or are mothballed. If we vanish, millions of CFC and HCFC automobile air conditioners, and millions more domestic and commercial refrigerators, refrigerated trucks and railroad cars, as well as home and industry air-cooling units, will all finally crack and give up the chlorofluorocarbonated ghost of a 20th-century idea that went very awry.

All will rise to the stratosphere, and the convalescing ozone layer will suffer a relapse. Since it won't happen all at once, with luck the illness will be chronic, not fatal. Otherwise, the plants and animals that remain in our wake will have to select for UV tolerance, or mutate their way through a barrage of electromagnetic radiation.

3. Tactical and Practical

Uranium-235, with a half-life of 704 million years, is a relatively insignificant fraction of natural uranium ore—barely .7 percent—but we humans have concentrated ("enriched") several thousand tons of it for use in reactors and bombs. To do that, we extract it from uranium ore, usually by chemically converting it to a gas compound, then spinning it in a centrifuge to separate the different atomic weights. This leaves behind far less potent ("depleted") U-238, whose half-life is 4.5 billion years: in the United States alone, there's at least a half-million tons of it.

One approach to what to do with some of it involves the fact that U-238 is an unusually dense metal. In recent decades it has proved useful, when alloyed with steel, for fashioning bullets that can pierce armor, including the walls of tanks.

With so much surplus depleted uranium lying around, this is far cheaper for U.S. and European armies than buying the non-radioactive alternative, tungsten, which is mainly found in China. Depleted uranium projectiles range from 25-millimeter bullet size to three-foot-long, 120-millimeter darts with their own internal propellants and stabilizing fins. Their use kindles outrage over human health issues, on both the firing and receiving end. Because depleted uranium ordnance bursts into flames when it strikes, it leaves a pile of ash. Depleted or not, there's enough concentrated U-238 in the bullet points that radioactivity in this debris can exceed 1,000 times the normal background level. After we're gone, the next archaeologists to appear may unearth arsenals of several million of these super-dense, modern versions of Clovis spear points. Not only will they look considerably more fearsome, but—possibly unbeknownst to their discoverers—they'll emit radiation for more years than the planet likely has left.

There are far hotter things than depleted uranium that will outlast us, whether we're gone tomorrow or 250,000 years from now. It's a big enough problem that we contemplate hollowing out entire mountains to store them. Thus far the United States has only one such site, in salt dome formations 2,000 feet below southeastern New Mexico, similar to the chemical-storage caverns below Houston. The Waste Isolation Pilot Plant, or WIPP, operating since 1999, is the boneyard for detritus from nuclear weapons and defense research. It can handle 6.2 million cubic feet of waste, the equivalent of about 156,000 55-gallon drums. In fact, much of the plutonium-drenched scrap it receives is packaged just that way.

WIPP isn't designed to store spent fuel from nuclear generating plants, which in the United States alone increases by 3,000 tons each year. It is a landfill only for so-called low- and midlevel waste—stuff like discarded weapons-assembly gloves, shoe coverings, and rags soaked in contaminated cleaning solvents used in fashioning nuclear bombs. It also holds the dismantled remains of machines used to build them, and even walls from rooms where that happened. All this arrives on shrink-wrapped pallets containing hot hunks of pipe, aluminum conduits, rubber, plastic, cellulose, and miles of wiring. After its first five years, WIPP was already more than 20 percent full.

Its contents come from two dozen high-security warrens across the country, such as the Hanford Nuclear Reservation in Washington, where plutonium for the Nagasaki bomb was made, and Los Alamos, New Mexico, where it was assembled. In 2000, large wildfires hit both sites. Official reports say that unburied radioactive wastes were protected—but in a world without firefighters, they won't be. Except for WIPP, all U.S. nuclear waste-storage containment is temporary. If it remains that way, fire will eventually breach it and send clouds of radioactive ash billowing across the continent, and possibly across the oceans.

The first site to begin shipping to WIPP was Rocky Flats, a defense facility on a foothills plateau 16 miles northwest of Denver. Until 1989, the United States made plutonium detonators for atomic weapons at Rocky Flats with somewhat less than a lawful regard for safety. For years, thousands of drums of cutting oil saturated with plutonium and uranium were stacked outside on bare ground. When someone finally noticed they were leaking, asphalt was poured over the evidence. Radioactive runoff at Rocky Flats frequently reached local streams; cement was swirled into radioactive sludge in absurd attempts to try to slow seepage from cracked evaporation ponds; and radiation periodically escaped into the air. A 1989 FBI raid finally closed the place. In the new millennium, after several billion dollars' worth of intensive cleanup and public relations, Rocky Flats was transmuted into a National Wildlife Refuge.

Simultaneously, similar alchemy was recasting the old Rocky Mountain Arsenal next to Denver International Airport. RMA was a chemical-weapons plant that made mustard and nerve gas, incendiary bombs, napalm—and during peacetime, insecticides; its core was once called the most contaminated square mile on Earth. After dozens of wintering bald eagles were found in its security buffer, feasting on the prodigious prairie dog population, it, too, became a National Wildlife Refuge. That required draining and sealing an Arsenal lake where ducks once died moments after landing, and where the bottoms of aluminum boats sent to fetch their carcasses rotted within a month. Although the plan is to treat and monitor toxic groundwater plumes for another century until they're considered safely diluted, today mule deer big as elk find asylum where humans once feared to tread.

A century, however, would make little difference to uranium and plutonium residues whose half-lives start at 24,000 years and keep going. The weapons-grade plutonium from Rocky Flats was shipped to South Carolina, whose governor was enjoined from lying in front of trucks to stop it. There, at the Savannah River Site's Defense Waste Processing

Facility, where two huge buildings ("reprocessing canyons") are so contaminated that no one knows how they might be decommissioned, high-level nuclear waste is now melted in furnaces with glass beads. When poured into stainless steel containers, it turns into solid blocks of radioactive glass.

This process, called vitrification, is also used in Europe. Glass being one of our simplest, most durable creations, these hot glass bricks may be among the longest-lasting of all human creations. However, in places like England's Windscale plant, scene of two nuclear accidents before it was finally closed, vitrified waste is stored in air-cooled facilities. One day, should power go off permanently, a chamber full of decaying, glass-embedded radioactive material would get steadily warmer, with shattering results.

The Rocky Flats asphalt where drums of radioactive oil spilled was also scraped and shipped to South Carolina, along with three feet of soil. More than half its 800 structures were razed, including the infamous "Infinity Room," where contamination levels rose higher than instruments could measure. Several buildings were mostly underground; after the removal of items like the glove boxes used to handle the shiny plutonium disks that triggered A-bombs, the basement floors were buried.

Atop them, a mix of native bluestem tall grass and side-oats grama grass has been planted to assure a habitat for resident elk, mink, mountain lion, and the threatened Prebel's meadow jumping mouse, which have impressively thrived in the plant's 6,000-acre security buffer despite the evil brewing at its center. Regardless of the grim business that went on here, these animals seem to be doing fine. However, while there are plans to monitor the human wildlife managers for radiation intake, a refuge official admits doing no genetic tests on the wildlife itself.

"We're looking at human hazards, not damage to species. Acceptable dose levels are based on 30-year career exposures. Most animals don't live that long."

Maybe not. But their genes do.

Anything at Rocky Flats too hard or too hot to move was covered with concrete and 20 feet of fill, and will remain off-limits to hikers in the wildlife preserve, though how they'll be deterred hasn't been decided. At

WIPP, where much of Rocky Flats ended up, the U.S. Department of Energy is legally required to dissuade anyone from coming too close for the next 10,000 years. After discussing the fact that human languages mutate so fast that they're almost unrecognizable after 500 or 600 years, it was decided to post warnings in seven of them anyway, plus pictures. These will be incised on 25-foot-high, 20-ton granite monuments and repeated on nine-inch disks of fired clay and aluminum oxide, randomly buried throughout the site. More-detailed information about the hazards below will go on the walls of three identical rooms, two of them also buried. The whole thing will be surrounded by a 33-foot-tall earthen berm a half-mile square, embedded with magnets and radar reflectors to give every possible signal to the future that *something* lurks below.

Whether who-or-whatever finds it someday can actually read, or heed, danger in those messages may be moot: the construction of this complex scarecrow to posterity isn't scheduled until decades from now, after WIPP is full. Also, after just five years, plutonium-239 was already noticed leaking from WIPP's exhaust shaft. Among the unpredictables is how all the irradiated plastic, cellulose, and radionuclides below will react as brine percolates through the salt formations, and as radioactive decay adds heat. For that reason, no radioactive liquids are allowed lest they volatilize, but many interred bottles and cans contain contaminated residues that will evaporate as temperatures rise. Head space is being left for buildup of hydrogen and methane, but whether it's enough, and whether WIPP's exhaust vent will function or clog, is the future's mystery.

4. Too Cheap to Meter

At the biggest U.S. nuclear plant, the 3.8-billion-watt Palo Verde Nuclear Generating Station in the desert west of Phoenix, water heated by a controlled atomic reaction turns to steam, which spins the three largest turbines General Electric ever manufactured. Most reactors worldwide function similarly; like Enrico Fermi's original atomic pile, all nuke plants use moveable, neutron-sopping cadmium rods to dampen or intensify the action.

In Palo Verde's three separate reactors, these dampers are interspersed among nearly 170,000 pencil-thin, 14-foot zirconium-alloy hollow rods stuffed end to end with uranium pellets that each contain as much power as a ton of coal. The rods are bunched into hundreds of assemblies; water flowing among them keeps things cool, and, as it vaporizes, it propels the steam turbines.

Together, the nearly cubical reactor cores, which sit in 45-foot-deep pools of turquoise water, weigh more than 500 tons. Each year, about 30 tons of their fuel is exhausted. Still packed inside the zirconium rods, this nuclear waste is removed by cranes to a flat-roofed building outside the containment dome, where it is submerged in a temporary holding pond that resembles a giant swimming pool, also 45 feet deep.

Since Palo Verde opened in 1986, its used fuel has been accumulating, because there's nowhere else to take it. In plants everywhere, spent fuel ponds have been re-racked to squeeze in thousands of more fuel assemblies. Together, the world's 441 functioning nuclear plants annually produce almost 13,000 tons of high-level nuclear scrap. In the United States, most plants have no more pool space, so until there's a permanent burial ground, waste-fuel rods are now mummified in "dry casks"—steel canisters clad in concrete from which the air and moisture have been sucked. At Palo Verde, where they've been used since 2002, these are stored vertically, and resemble giant thermos bottles.

Every country has plans to permanently entomb the stuff. Every country also has citizens terrified of events like earthquakes that could unseal buried waste, and of the chance that some truck carrying it will have a wreck or be hijacked en route to the landfill.

In the meantime, used nuclear fuel, some of it decades old, languishes in holding tanks. Oddly, it is up to a million times more radioactive than when it was fresh. While in the reactor, it began mutating into elements heavier than enriched uranium, such as isotopes of plutonium and americium. That process continues in the waste dumps, where used hot rods exchange neutrons and expel alpha and beta particles, gamma rays, and heat.

If humans suddenly departed, before long the water in the cooling ponds would boil and evaporate away—rather quickly in the Arizona desert. As the used fuel in the storage racks is exposed to air, its heat would ignite the cladding of the fuel rods, and radioactive fire would break out. At Palo Verde, like other reactors, the spent-fuels building was intended to be temporary, not a tomb, and its masonry roof is more similar to a big-box discount store's than to the reactor's pre-stressed containment

Reloading nuclear fuel: Unit 3, Palo Verde Nuclear Generating Station.

dome. Such a roof wouldn't last long with a radioactive fire cooking below it, and much contamination would escape. But that wouldn't be the biggest problem.

Resembling giant enoki mushrooms, Palo Verde's great steam columns rise a mile over the desert creosote flats, each consisting of 15,000 gallons of water evaporated per minute to cool Palo Verde's three fission reactors. (As Palo Verde is the only U.S. plant not on a river, bay, or seacoast, the water is recycled Phoenix effluent.) With 2,000 employees to keep pumps from sticking, gaskets from leaking, and filters back-washed, the plant is a town big enough to have its own police and fire departments.

Suppose its inhabitants had to evacuate. Suppose they had enough advance warning to shut down by jamming all the moderating rods into each reactor core to stop the reaction and cease generating electricity. Once Palo Verde was unmanned, its connection to the power grid would automatically be cut. Emergency generators with a seven-day diesel supply would kick in to keep coolant water circulating, because even if fission in the core stopped, uranium would continue to decay, generating about 7 percent as much heat as an active reactor. That heat would be enough to keep pressurizing the cooling water looping through the reactor core. At times, a relief valve would open to release overheating water, then close again when the pressure dropped. But the heat and pressure would build again, and the relief valve would have to repeat its cycle.

At some point, it becomes a question of whether the water supply is depleted, a valve sticks, or the diesel pumps cut out first. In any case, cooling water will cease being replenished. Meanwhile, the uranium fuel, which takes 704 million years to lose just half its radioactivity, is still hot. It keeps boiling off the 45 feet of water in which it sits. In a few weeks at the most, the top of the reactor core will be exposed, and the meltdown will begin.

If everyone had vanished or fled with the plant still producing electricity, it would keep running until any one of thousands of parts monitored daily by maintenance personnel failed. A failure should automatically trigger a shutdown; if it didn't, the meltdown might occur quite quickly. In 1979 something similar happened at Pennsylvania's Three Mile Island Plant when a valve stuck open. Within two hours and

15 minutes, the top of the core was exposed and turned into lava. As it flowed to the bottom of the reactor vessel, it started burning through six inches of carbon steel.

It was a third of the way through before anyone realized. Had no one discovered the emergency, it would have dropped into the basement, and 5,000°F molten lava would have hit nearly three feet of water flooded from the stuck valve, and exploded.

Nuclear reactors have far less concentrated fissionable material than nuclear bombs, so this would have been a steam explosion, not a nuclear explosion. But reactor containment domes aren't designed for steam explosions; as its doors and seams blow out, a rush of incoming air would immediately ignite anything handy.

If a reactor was near the end of its 18-month refueling cycle, a meltdown to lava would be more likely, because months of decay build up considerable heat. If the fuel was newer, the outcome might be less catastrophic, though ultimately just as deadly. Lower heat might cause a fire instead of a meltdown. If combustion gases shattered the fuel rods before they turned to liquid, uranium pellets would scatter, releasing their radioactivity inside the containment dome, which would fill with contaminated smoke.

Containment domes are not built with zero leakage. With power off and its cooling system gone, heat from fire and fuel decay would force radioactivity out gaps around seals and vents. As materials weathered, more cracks would form, seeping poison, until the weakened concrete gave way and radiation gushed forth.

If everyone on Earth disappeared, 441 nuclear plants, several with multiple reactors, would briefly run on autopilot until, one by one, they overheated. As refueling schedules are usually staggered so that some reactors generate while others are down, possibly half would burn, and the rest would melt. Either way, the spilling of radioactivity into the air, and into nearby bodies of water, would be formidable, and it would last, in the case of enriched uranium, into geologic time.

Those melted cores that flow to the reactor floors would not, as some believe, bore through the Earth and out the other side, emerging in China like poisonous volcanoes. As the radioactive lava melds with the surrounding steel and concrete, it would finally cool—if that's the term for a lump of slag that would remain mortally hot thereafter.

That is unfortunate, because deep self-interment would be a blessing to whatever life remained on the surface. Instead, what briefly was an exquisitely machined technological array would have congealed into a deadly, dull metallic blob: a tombstone to the intellect that created it—and, for thousands of years thereafter, to innocent nonhuman victims that approach too closely.

5. Hot Living

They began approaching within a year. Chernobyl's birds disappeared in the firestorm when Reactor Number Four blew that April, their nest building barely begun. Until it detonated, Chernobyl was almost halfway to becoming the biggest nuclear complex on Earth, with a dozen one-megawatt reactors. Then, one night in 1986, a collision of operator and design mistakes achieved a kind of critical mass of human error. The explosion, although not nuclear—only one building was damaged—broadcast the innards of a nuclear reactor over the landscape and into the sky, amid an immense cloud of radioactive steam from the evaporated coolant. To Russian and Ukrainian scientists that week, frantically sampling to track radioactive plumes through the soil and aquifers, the silence of a birdless world was unnerving.

But the following spring the birds were back, and they've stayed. To watch barn swallows zip naked around the carcass of the hot reactor is discombobulating, especially when you are swaddled in layers of wool and hooded canvas coveralls to block alpha particles, with a surgical cap and mask to keep plutonium dust from your hair and lungs. You want them to fly away, fast and far. At the same time, it's mesmerizing that they're here. It seems so normal, as if apocalypse has turned out to be not so bad after all. The worst happens, and life still goes on.

Life goes on, but the baseline has changed. A number of swallows hatch with patches of albino feathers. They eat insects, fledge, and migrate normally. But the following spring, no white-flecked birds return. Were they too genetically deficient to make the winter circuit to southern Africa? Does their distinctive coloring make them unappealing to potential mates, or too noticeable to predators?

In the aftermath of Chernobyl's explosion and fire, coal miners and subway crews tunneled underneath Number Four's basement and poured a second concrete slab to stop the core from reaching groundwater. This probably was unnecessary, as the meltdown was over, having ended in a 200-ton puddle of frozen, murderous ooze at the bottom of the unit. During the two weeks it took to dig, workers were handed bottles of vodka, which, they were told, would inoculate them against radiation sickness. It didn't.

At the same time, construction began on a containment housing, something that all Soviet RMBK reactors like Chernobyl lacked, because they could be refueled faster without one. By then, hundreds of tons of hot fuel had already blown onto the roofs of adjacent reactors, along with 100 to 300 times the radiation released in the 1945 bombing of Hiroshima. Within seven years, radioactivity had eaten so many holes in the hastily built, hulking, gray five-story concrete shell, already patched and caulked like the hull of a rusting scow, that birds, rodents, and insects were nesting inside it. Rain had leaked in, and no one knew what vile brews steeped in puddles of animal droppings and warm, irradiated water.

The Zone of Alienation, a 30-kilometer-radius evacuated circle around the plant, has become the world's biggest nuclear-waste dump. The millions of tons of buried hot refuse include an entire pine forest that died within days of the blast, which couldn't be burned because its smoke would have been lethal. The 10-kilometer radius around ground zero, the plutonium zone, is even more restricted. Any vehicles and machinery that worked there on the cleanup, such as the giant cranes towering over the sarcophagus, are too radioactive to leave.

Yet skylarks perch on their hot steel arms, singing. Just north of the ruined reactor, pines that have re-sprouted branch in elongated, irregular runs, with needles of various lengths. Still, they're alive and green. Beyond them, by the early 1990s, forests that survived had filled with radioactive roe deer and wild boars. Then moose arrived, and lynx and wolves followed.

Dikes have slowed radioactive water, but not stopped it from reaching the nearby Pripyat River and, farther downstream, Kiev's drinking supply. A railroad bridge leading to Pripyat, the company town where 50,000 were evacuated—some not quickly enough to keep radioactive iodine from ruining their thyroids—is still too hot to cross. Four miles south, though, you can stand above the river in one of the best birding areas to-day in Europe, watching marsh hawks, black terns, wagtails, golden and white-tailed eagles, and rare black storks sail past dead cooling towers.

In Pripyat, an unlovely cluster of concrete 1970s high-rises, returning poplars, purple asters, and lilacs have split the pavement and invaded buildings. Unused asphalt streets sport a coat of moss. In surrounding villages, vacant except for a few aged peasants permitted to live out their shortened days here, stucco peels from brick houses engulfed by untrimmed shrubbery. Cottages of hewn timbers have lost roof tiles to tangles of wild grapevines and even birch saplings.

Just beyond the river is Belarus; the radiation, of course, stopped for no border. During the five-day reactor fire, the Soviet Union seeded clouds headed east so that contaminated rain wouldn't reach Moscow. In-stead, it drenched the USSR's richest breadbasket, 100 miles from Cher-nobyl at the intersection of Ukraine, Belarus, and western Russia's Novozybkov region. Except for the 10-kilometer zone around the reactor, no other place received so much radiation—a fact concealed by the Soviet government lest national food panic erupt. Three years later, when re-searchers discovered the truth, most of Novozybkov was also evacuated, leaving fallow vast collective grain and potato fields.

The fallout, mainly cesium-137 and strontium-90, by-products of uranium fission with 30-year half-lives, will significantly irradiate Novozybkov's soils and food chain until at least AD 2135. Until then, nothing here is safe to eat, for either humans or animals. What "safe" means is wildly debated. Estimates of the number of people who will die from cancer or blood and respiratory diseases due to Chernobyl range from 4,000 to 100,000. The lower figure comes from the International Atomic Energy Agency, whose credibility is tinged by its dual role as both the world's atomic watchdog agency and the nuclear power industry's trade association. The higher numbers are invoked by public health and cancer researchers and by environmental groups like Greenpeace Interna-tional, all insisting that it's too early to know, because radiation's effects accumulate over time.

———————

Whatever the correct measure of human mortality may be, it applies to other life-forms as well, and in a world without humans the plants and animals we leave behind will have to deal with many more Chernobyls. Little is still known about the extent of genetic harm this disaster unleashed: genetically damaged mutants usually fall to predators before scientists can count them. However, studies suggest that the survival rate of Chernobyl swallows is significantly lower than that of returning migrants of the same species elsewhere in Europe.

"The worst-case scenario," remarks University of South Carolina biologist Tim Mousseau, who visits here often, "is that we might see extinction of a species: a mutational meltdown."

"Typical human activity is more devastating to biodiversity and abundance of local flora and fauna than the worst nuclear power plant disaster," dourly observe radioecologists Robert Baker, of Texas Tech University, and Ronald Chesser, of the University of Georgia's Savannah River Ecology Laboratory, in another study. Baker and Chesser have documented mutations in the cells of voles in Chernobyl's hot zone. Other research on Chernobyl's voles reveals that, like its swallows, the life-spans of these rodents are also shorter than those of the same species elsewhere. However, they seem to compensate by sexually maturing and bearing offspring earlier, so their population hasn't declined.

If so, nature may be speeding up selection, upping the chances that somewhere in the new generation of young voles will be individuals with increased tolerance to radiation. In other words, mutations—but stronger ones, evolved to a stressed, changing environment.

Disarmed by the unexpected beauty of Chernobyl's irradiated lands, humans have even tried to encourage nature's hopeful bravado by reintroducing a legendary beast not seen in these parts for centuries: bison, brought from Belarus's Belovezhskaya Pushcha, the relic European forest it shares with Poland's Białowieża Puszcza. So far, they're grazing peacefully, even nibbling the bitter namesake wormwood—*chornobyl* in Ukrainian.

Whether their genes will survive the radioactive challenge will only be known after many generations. There may be more challenges: A new sarcophagus to enclose the old, useless one, isn't guaranteed to last, either.

Eventually, when its roof blows away, radioactive rainwater inside and in adjacent cooling ponds could evaporate, leaving a new lode of radioactive dust for the burgeoning Chernobyl menagerie to inhale.

After the explosion, the radionuclide count was high enough in Scandinavia that reindeer were sacrificed rather than eaten. Tea plantations in Turkey were so uniformly dosed that Turkish tea bags were used in Ukraine to calibrate dosimeters. If, in our wake, we leave the cooling ponds of 441 nuclear plants around the world to dry and their reactor cores to melt and burn, the clouds enshrouding the planet will be far more insidious.

Meanwhile, we are still here. Not just animals but people too have crept back into Chernobyl's and Novozybkov's contaminated zones. Technically, they're illegal squatters, but authorities don't try very hard to dissuade the desperate or needy from gravitating to empty places that smell so fresh and look so clean, as long as no one checks those dosimeters. Most of them aren't simply seeking free real estate. Like the swallows who returned, they come because they were here before. Tainted or not, it's something precious and irreplaceable, even worth the risk of a shorter life.

It's their home.

CHAPTER 16

⚜

Our Geologic Record

1. Holes

ONE OF THE largest, and probably longest-lasting, relics of human existence after we're gone is also one of the youngest. As the gyrfalcon flies, it lies 180 miles northeast of Yellowknife, Northwest Territories, Canada. If you flew over it today, it would be the very round hole half a mile wide and 1,000 feet deep. There are many huge holes here. This is the dry one.

Though, within a century, the rest may be, too. North of the 60th parallel, Canada contains more lakes than the rest of the world combined. Nearly half of Northwest Territories isn't land at all, but water. Here, ice ages gouged cavities into which icebergs dropped when the glaciers retreated. When they melted, these earthen kettles filled with fossil water, leaving countless mirrors that sequin the tundra. Yet the resemblance to an immense sponge is misleading: because evaporation slows in cold climates, little more precipitation falls here than in the Sahara. Now, as the permafrost thaws around these kettles, glacial water held in place by frozen soil for thousands of years is seeping away.

Should northern Canada's sponge dry out, that would also be a human legacy. For now, the hole in question and two recent smaller ones nearby comprise Ekati, Canada's first diamond mine. Since 1998, a parade of 240-ton trucks with 11-foot tires, owned by BHP Billiton Diamonds, Inc., has lugged more than 10,000 tons of ore to a crusher 24 hours a day, 365 days a year, even at −60°F. The daily yield is a handful of gem-quality diamonds, worth well over $1 million.

They're found in volcanic tubes that formed more than 50 million years ago, when magma bearing pure, crystallized carbon pushed up from deep below the surrounding granite. Even rarer than these diamonds, however, is what fell into the craters left by these lava pipes. This was the Eocene, when today's lichen-covered tundra was a coniferous forest. The first trees that toppled in must have burned, but as everything cooled, others were buried in fine ash. Sealed away from air, then preserved by cold Arctic dryness, the fir and redwood trunks that diamond miners discovered weren't even fossilized, but simply wood: intact, 52-million-year-old lignin and cellulose, dating back to when mammals were just expanding into niches vacated by dinosaurs.

One of the oldest mammal species on Earth still lives there, a Pleistocene relic that managed to survive because it was extraordinarily equipped to brave weather that ice-age humans preferred to escape. The chestnut pelage of the musk ox is the warmest organic fiber known, with eight times the insulating factor of sheep's wool. Known in Inuit as *qiviut,* it renders musk oxen so impervious to cold that they're literally invisible to infrared satellite cameras used to track caribou herds. Yet *qiviut* was nearly their downfall in the early 20th century, when they were all but exterminated by hunters who sold their hides in Europe for carriage robes.

Today, the few thousand that remain are protected, and the only legally harvested wild *qiviut* comes from wisps found clinging to tundra vegetation, a laborious undertaking that contributes to the $400 price a sweater made from ultrasoft musk ox fleece can command. If the Arctic becomes progressively balmier, however, *qiviut* may again be this species' undoing—although, if humans (or at least their noisy carbon emitters) vanished, the musk ox might still get a break from the heat.

If too much of the permafrost itself is undone, it would thaw deeply buried ice that forms crystalline cages around methane molecules. An estimated 400 billion tons of these frozen methane deposits, known as clathrates, lie a few thousand feet beneath the tundra, and even more are found beneath the world's oceans. All that very-deep-freeze natural gas, estimated to at least equal all known conventional gas and oil reserves, is both enticing and frightening. Because it's so dispersed, no one has come up with an economical way to mine it. Because there's so much of it, if it all floats away once the ice cages melt, that much methane might ratchet global warming to levels unknown since the Permian extinction, 250 million years ago.

For now, until something cheaper and cleaner comes along, the only other still-abundant source of fossil fuel we can count on will leave an even bigger signature on the surface than a mere open-pit diamond mine—or copper, iron, or uranium mines, for that matter. Long after those fill with water or with their own windblown tailings, this one is good to endure a few more million years.

2. Heights

"It can only be appreciated—if that's the word—from above," says Susan Lapis, a cheerful redheaded pilot who volunteers for the North Carolina–based nonprofit organization SouthWings. From the window of her single-engine, red, white, and blue Cessna 182, you look down at a world sliced as flat as any mile-high ice sheet ever managed. Only this time, the glacier was us, and the world was once West Virginia.

Or Virginia, Kentucky, or Tennessee, because several million acres of Appalachia in all those states now look identically amputated, sheared away by coal companies that, in the 1970s, discovered a trick cheaper than tunneling or even strip-mining: just pulverize the entire top third of a mountain, sluice out the coal with a few million gallons of water, push what's left over the side, and blast again.

Not even the Amazon laid bare rivals the shock of this planar void. In every direction, it's simply gone. Grids of white dots—the next round of dynamite charges—provide the only remaining texture on naked plateaus that were once vertical, verdant heights. Demand for coal has been so ferocious—100 tons extracted every two seconds—that often there hasn't even been time to log here: oaks, hickories, magnolias, and black cherry hardwoods have been bulldozed into the hollows, to be buried by a former Allegheny mountain of rubble—"the overburden."

In West Virginia alone, 1,000 miles of streams flowing through those hollows have been buried as well. Water finds a way, of course, but as it pushes through tailings for the next several thousand years, it will emerge with more than the normal concentration of heavy metals. Yet even given

Mountaintop removal, West Virginia.
PHOTO BY V. STOCKMAN, OVEC/SOUTHWINGS.

projected world energy demand, industry geologists—and industry opponents—believe that deposits in the United States, China, and Australia contain about 600 years' worth of coal. By mining this way, they can get at much more of it, much faster.

If energy-drunk humans were gone tomorrow, all that coal would remain in the ground until the end of time on Earth. If we're around at least a few more decades, however, a lot of it won't, because we'll dig it up and burn it. But if an unlikely plan goes extremely well, one of coal power's most problematic by-products may end up sealed away once more beneath the surface, creating yet another human legacy to the far future.

The by-product is carbon dioxide, which a burgeoning consensus of humanity agrees probably should not be stored in the atmosphere. The plan, which is attracting growing attention—especially from industry boosters of an oxymoron born of recombinant public relations: "clean coal"—is to capture CO_2 before it leaves the smokestack of coal-fired electrical plants, stuff it underground, and keep it there. Forever.

It would work like this: Pressurized CO_2 would be injected into saline aquifers that, in much of the world, lie under impermeable caprock at depths of 1,000 to 8,000 feet. There, supposedly, the CO_2 would go into solution, forming mild carbonic acid—like salty Perrier. Gradually, the carbonic acid would react with surrounding rocks, which would dissolve and slowly precipitate out as dolomite and limestone, locking the greenhouse gas in stone.

Each year since 1996, Norway's Statoil has sequestered 1 million tons of carbon dioxide in a saline formation under the North Sea. In Alberta, CO_2 is being sequestered in abandoned gas wells. Back in the 1970s, then federal attorney David Hawkins joined in discussions with semioticians about how people 10,000 years hence might be alerted to buried nuclear wastes at what today is New Mexico's WIPP site. Now, as director of the Climate Center of the Natural Resources Defense Council, he contemplates how to tell the future not to drill into sequestered reservoirs of invisible gases we might sweep under the rug, lest they unexpectedly burp to the surface.

Aside from the expense of drilling enough holes to capture, pressurize, and inject the CO_2 from every industrial and power plant on Earth, a big concern is that hard-to-detect leaks of even $1/10$ of 1 percent would eventually add up to the amount we're pumping into the air today—and the future wouldn't realize it. But given the choice, Hawkins would rather try containing carbon than plutonium.

"We know that nature can engineer leak-free gas storage: there's been methane trapped for millions of years. The question is, can humans?"

3. Archaeological Interlude

We tear down mountains, and unwittingly build hills.

Forty minutes northeast of the city of Flores on northern Guatemala's Lake Petén Ixta, a paved tourist road arrives at the ruins of Tikal, the largest Classic Mayan site, its white temples rising 230 feet above the jungle floor.

In the opposite direction, until recent improvements haived travel time, the rutted road southwest from Flores took three miserable hours, ending at the scruffy outpost of Sayaxché, where an army machine gun placement perched atop a Mayan pyramid.

Sayaxché is on the Río Pasión—the Passion River—which lolls through the western Petén province to the confluence of the rivers Usamacinta and Salinas, together forming Guatemala's border with Mexico. The Pasión was once a major trade route for jade, fine pottery, quetzal feathers, and jaguar skins. More recently, commerce includes contraband mahogany and cedar logs, opium from Guatemalan highland poppies, and looted Mayan artifacts. During the early 1990s, motor-driven wooden launches on a sluggish Pasión tributary, the Riochuelo Petexbatún, also carried quantities of two modest items that in the Petén are veritable luxuries: corrugated zinc roofing and cases of Spam.

Both were destined for the base camp that Vanderbilt University's Arthur Demarest built in a jungle clearing out of mahogany planks for one of the biggest archaeological excavations in history, to solve one of our biggest mysteries: the disappearance of Mayan civilization.

How can we even contemplate a world without us? Fantasies of space aliens with death rays are, well, fantasies. To imagine our big, overwhelming civilization *really* ending—and ending up forgotten under layers of dirt and earthworms—is as hard for us as picturing the edge of the universe.

The Maya, however, were real. Their world had seemed destined to thrive forever, and, at its zenith, it was far more entrenched than ours. For at least 1,600 years, about 6 million Maya lived in what in some ways resembled southern California—a flourishing megalopolis of city-states, with few breaks between overlapping suburbs across a lowland that today comprises northern Guatemala, Belize, and Mexico's Yucatán Peninsula. Their commanding architecture, and their astronomy, mathematics, and literature, would have humbled the achievements of contemporaries in Europe. Equally striking, and far less understood, is how so many could inhabit a tropical rain forest. For centuries, they raised their food and families in the same fragile environment that today is quickly being devastated by relatively few hungry squatters.

What has baffled archaeologists even more, however, is the Maya's spectacular, sudden collapse. Beginning in the eighth century AD, within just 100 years lowland Mayan civilization vanished. In most of the Yucatán, only scattered remnants of the population remained.

Northern Guatemala's Petén province was virtually a world without people. Rainforest vegetation soon overran the ball courts and plazas, enshrouding tall pyramids. Not for 1,000 years would the world again be aware of their existence.

But the Earth holds ghosts, even of entire nations. Archaeologist Arthur Demarest, a stocky, thick-moustached Louisiana Cajun, declined a Harvard chair because Vanderbilt offered him a chance to exhume this one. During his graduate fieldwork in El Salvador, Demarest raced to salvage a bit of the ancient record from a forthcoming dam that displaced thousands, converting many of them to guerrillas. When three of his workers were accused of being terrorists, he pleaded to officials who let them go, but they were assassinated anyway.

In his first years in Guatemala, guerrillas and the army stalked each other within a few kilometers of his digs, catching in their crossfire people who still speak languages derived from the hieroglyphics his team was decoding.

"Indiana Jones swashbuckled through a mythical, generic Third World of swarthy people with threatening, incomprehensible ways, defeating them with American heroics and seizing their treasures," he says, mopping his thick black hair. "He would have lasted five seconds here. Archaeology isn't about glittery objects—it's about their context. We're part of the context. It's our workers whose fields are burning, it's their children who have malaria. We come to study ancient civilization, but we end up learning about now."

By Coleman lantern, he writes through the humid night to the rumble of howler monkeys, piecing together how, over nearly two millennia, the Maya evolved a means of resolving discord between nations without destroying each other's societies in the process. But then something went wrong. Famine, drought, epidemics, overpopulation, and environmental plunder have been blamed for the Maya's downfall—yet for each, arguments exist against liquidation on such a massive scale. No relics reveal an alien invasion. Often extolled as an exemplary stable and peaceable people, the Maya seemed least likely to overreach and be devoured by their own greed.

However, in the steamy Petén, it appears that is exactly what happened—and that the path to their catastrophe seems familiar.

The trek from the Riochuelo Petexbatún to Dos Pilas, the first of seven major sites that Demarest's team uncovered, passes for hours through mosquito-rich stretches of strangler vines and *palmilla* thickets, then finally climbs a steep escarpment. In remaining groves still unplundered by timber poachers, giant cedars, ceibas, chicle-bearing sapodillas, mahoganies, and breadnut trees rise from the thin tropical soils capping the Petén limestone. Along the escarpment's ragged edge, the Maya built cities that Arthur Demarest's archaeologists have determined once formed an interlocking kingdom called Petexbatún. Today, what appear to be hills and ridges are actually pyramids and walls, built from chunks of local limestone hewn with chert adzes, now disguised by soil and a mature rain forest.

The jungle surrounding Dos Pilas, filled with clacking toucans and parrots, was so dense that after it was discovered in the 1950s, 17 years passed before anyone noticed that a nearby hill was actually a 220-foot pyramid. In fact, to the Maya, pyramids re-created mountains, and their carved monoliths, called stelae, were stone representations of trees. The dot-and-bar code glyphs carved into the stelae unearthed around Dos Pilas tell that, about AD 700, its *k'uhul ajaw*—divine lord—began to break the rules of restrained conflict and started usurping neighboring Petexbatún city-states.

A mossy stela shows him in full headdress, holding a shield, standing on the back of a bound human captive. Before society began unraveling, Classic Mayan wars were often keyed to astrological cycles, and at first impression they might seem singularly grisly. A male of an opposing royal family would be captured and paraded in humiliation, sometimes for years. Eventually, his heart would be ripped out, or he would be decapitated, or tortured to death—at Dos Pilas, one victim was tightly rolled and bound, and then used for a game on the ceremonial ball court until his back was broken.

"And yet," Demarest notes, "there was relatively no societal trauma, no destruction of fields or buildings, or territories taken. The cost of Classic Mayan ritual war was minimal. It was a way of maintaining peace through constant, low-grade warfare that released tensions between leaders without endangering the landscape."

The landscape was a working equilibrium between wilderness and artifice. On hillsides, Mayan walls of tightly packed cobbles trapped rich humus from runoff water for cultivation terraces, now lost beneath a millennium of alluvium. Along lakes and rivers, Mayas dug ditches to drain swamps, and by heaping the soil they removed, they created fertile raised fields. Mostly, though, they mimicked the rain forest, providing layered shade for diverse crops. Rows of corn and beans would shelter a ground cover of melons and squash; fruit trees, in turn, shielded them, and protective patches of the forest itself would be left among fields. Partly, it was a happy accident: without chain saws, they had to leave the biggest trees.

That is exactly what hasn't happened in nearby modern squatter villages, along logging roads where flatbed trailers carry away cedar and mahogany. The settlers, Mayan-Kekchi-speaking refugees from the highlands, fled counterinsurgency attacks that killed thousands of Guatemalan peasants during the 1980s. Because slash-and-burn rotations used in volcanic mountains prove calamitous in rain forests, these people were soon surrounded by expanding wastelands yielding only stunted ears of maize. To keep them from looting all his sites, Demarest budgets for doctors and jobs for locals.

The Maya's political and agricultural system functioned for centuries throughout the lowlands, until it began to break down at Dos Pilas. During the eighth century, new stelae began appearing, with the creative flair of individual sculptors supplanted by uniform, military social realism. Gaudy hieroglyphics incised on each tier of an elaborate temple staircase record victories over Tikal and other centers, whose glyphs were replaced by those of Dos Pilas. For the first time, land was being conquered.

Strategically parlaying alliances with other rival Mayan city-states, Dos Pilas metastasized into an aggressive international power whose influence advanced up the Río Pasión's valley to today's Mexican border. Its artisans planted stelae portraying a Dos Pilas *k'uhul ajaw* resplendent in jaguar-skin boots, with a naked, vanquished king crushed under his feet. Dos Pilas's rulers amassed fabulous wealth. In caves where no human had been for 1,000 years, Demarest and his colleagues found where they hoarded hundreds of ornate polychromed pots containing jade, flint, and the remains of sacrificed humans. In tombs the archaeologists exhumed, royalty were buried with their mouths full of jade.

By AD 760, the domain they and their allies controlled encompassed more than three times a normal Classic Mayan kingdom. But they now barricaded their cities with palisades, spending much of their reign behind walls. A remarkable discovery bears witness to the end of Dos Pilas itself. Following an unexpected defeat, no more self-aggrandizing monuments were built. Instead, peasants who lived in concentric rings of fields around the city fled their houses, erecting a squatters' village in the middle of the ceremonial plaza. The degree of their panic is preserved in the bulwark they threw up around their compound, made from facing materials ripped from a *k'uhul ajaw*'s tomb and from the principal palace, whose corbeled temple was demolished and added to the rubble. It was the equivalent of tearing down the Washington Monument and Lincoln Memorial to fortify a tent city on the Capitol Mall. The desecration heightened as the wall ran right over the top of structures, including the triumphant hieroglyphic staircase.

Had these crude placements possibly occurred much later? That question was answered by the facing stones they found in direct contact with the stairs, with no intervening soil. The citizens of Dos Pilas, either beyond reverence for, or thoroughly outraged by, the memory of their greedy former rulers, did this themselves. They buried the magnificent carved hieroglyphic staircase so deeply that no one knew it existed until a Vanderbilt graduate student uncovered it 1,200 years later.

Did mounting population exhaust the land, tempting Petexbatún rulers to seize their neighbors' property, leading to a cycle of response that spiraled into cataclysmic war? If anything, Demarest believes, it was the other way around: An unleashed lust for wealth and power turned them into aggressors, resulting in reprisals that required their cities to abandon vulnerable outlying fields and intensify production closer to home, eventually pushing land beyond its tolerance.

"Society had evolved too many elites, all demanding exotic baubles." He describes a culture wobbling under the weight of an excess of nobles, all needing quetzal feathers, jade, obsidian, fine chert, custom polychrome, fancy corbeled roofs, and animal furs. Nobility is expensive, nonproductive, and parasitic, siphoning away too much of society's energy to satisfy its frivolous cravings.

"Too many heirs wanted thrones, or needed some ritual bloodletting to confirm their stature. So dynastic warfare heightened." As more temples need building, the higher caloric demand on workers requires more food production, he explains. Population rises to insure enough food-producers. War itself often increases population—as it did in the Aztec, Incan, and Chinese empires—because rulers require cannon fodder.

Stakes rise, trade is disrupted, and population concentrates—lethal in a rain forest. There is dwindling investment in long-term crops that maintain diversity. Refugees living behind defensive walls farm only adjacent areas, inviting ecological disaster. Their confidence in leaders who once seemed all-knowing, but are obsessed with selfish, short-term goals, declines with the quality of life. People lose faith. Ritual activity ceases. They abandon centers.

A ruin at nearby Lake Petexbatún, on a peninsula called Punta de Chimino, turned out to be the fortress city of the last Dos Pilas *k'uhul ajaw*. The peninsula had been severed from the mainland by three moats, one cutting so deeply into bedrock that approximately three times the energy required to build the city itself was expended to dig it. "That's the equivalent," Demarest observes, "of spending 75 percent of a nation's budget on defense."

It was a desperate society that had lost control. The spear points the archaeologists discovered embedded in the fortress walls—including on the inside—testify to the fate of whoever ended up cornered on Punta de Chimino. Its monuments were soon eaten by the forest: in a world relieved of its humans, man's attempts to make his own mountains quickly melt back into the ground.

"When you examine societies just as self-confident as ours that unraveled and were eventually swallowed by the jungle," says Arthur Demarest, "you see that the balance between ecology and society is exquisitely delicate. If something throws that off, it all can end."

He stoops, picks up a sherd from the moist ground. "Two thousand years later, someone will be squinting over the fragments, trying to find out what went wrong."

4. Metamorphosis

From a wooden crate on the floor of his office at the Smithsonian's National Museum of Natural History, paleobiology curator Doug Erwin pulls an eight-inch chunk of limestone he found in a phosphate mine between Nanjing and Shanghai, south of China's Yangtze River. He shows the blackish bottom half, replete with fossilized protozoa, plankton, univalves, bivalves, cephalopods, and corals. "Life here was good." He points to the faint whitish line of ash that separates it from the dull gray upper half. "Life here got really bad." He shrugs.

"It then took a long time for life to get better."

It took dozens of Chinese paleontologists 20 years of examining such rocks to determine that the faint white line represents the Permian Extinction. By analyzing zircon crystals infused in tiny glassy and metallic globules embedded in it, Erwin and MIT geologist Sam Bowring precisely dated that line to 252 million years ago. The black limestone lying below it is a frozen snapshot of the rich coastal life that had surrounded a single giant continent filled with trees, crawling and flying insects, amphibians, and early carnivorous reptiles.

"Then," says Erwin, nodding, "95 percent of everything alive on the planet was wiped out. It was actually a thoroughly good idea."

Sandy-haired Doug Erwin looks improbably boyish for so distinguished a scientist. When he dismisses life-on-Earth's closest brush with total annihilation, however, his smile isn't flippant but thoughtful—the result of decades of poring over West Texas mountains, old Chinese quarries, and ravines in Namibia and South Africa to puzzle out what exactly happened. He still doesn't know for sure. A million-year-long volcanic eruption through enormous coal deposits in Siberia (then part of Pangea, the single supercontinent) flooded the land with so much basalt magma— in places, it was more than three miles thick—that CO_2 from vaporized coal may have glutted the atmosphere and sulfuric acid may have rained from the skies. The coup de grâce may have been an asteroid even bigger than the one that did in the dinosaurs much later; it apparently collided with the piece of Pangea we now call Antarctica.

Whatever it was, over the next few million years, the most common vertebrate was a microscopic-toothed worm. Even insects suffered a mass collapse. This was a good idea?

"Sure. It made way for the Mesozoic Era. The Paleozoic had been around for nearly 400 million years. It was fine, but it was time to try something new."

Following the Permian's fiery end, the few survivors had little competition. One of them, a half-dollar-sized, scalloplike clam called *claraia*, became so abundant that its fossils today literally pave rocks in China, southern Utah, and northern Italy. But within 4 million years, they and most other bivalves and snails that boomed after the mass extinction died out themselves. They were victims of more mobile opportunists such as crabs, who'd had minor roles in the old ecosystem, but suddenly—at least by the geologic clock—had a chance at creating new niches in a fresh new system. All it took was evolving a claw to crack open mollusks that couldn't flee.

The world took off in a different direction—one characterized by active predators—that went from near nothingness to the lush kingdom of dinosaurs. While that was happening, the supercontinent split into pieces that gradually dispersed around the globe. When, after 150 million more years, that other asteroid hit what is now Mexico's Yucatán Peninsula, and dinosaurs proved too big to hide or adapt, it was time to start over yet again. This time, another agile minor character, a vertebrate called *Mammalia,* saw a chance to make its move.

Might the current explosion of extinctions—invariably pointing to a sole cause, and not an asteroid this time—suggest that a certain dominant mammal's turn may be coming to an end? Is geologic history striking again? Doug Erwin, the extinction expert, works on such a vast timescale that the few million years of our *Homo* species' life span is almost too short for him to contemplate. Again, he shrugs.

"Humans are going extinct eventually. Everything has, so far. It's like death: there's no reason to think we're any different. But life will continue. It may be microbial life at first. Or centipedes running around. Then life will get better and go on, whether we're here or not. I figure it's interesting to be here now," he says. "I'm not going to get all upset about it."

If humans do stick around, University of Washington paleontologist Peter Ward predicts that agricultural land will become the biggest habitat on Earth. The future world, he believes, will be dominated by whatever evolves from the handful of plants and animals we've domesticated for food, work, raw materials, and companionship.

But if humans were to go tomorrow, enough wild predators currently remain to out-compete or gobble most of our domestic animals, though a few feral exceptions have proved impressively resilient. The escaped wild horses and burros of the American Great Basin and Sonoran Desert essentially have replaced equine species lost at the end of the Pleistocene. Dingoes, which polished off Australia's last marsupial carnivores, have been that country's top predator for so long that many down under don't realize that these canines were originally companions to Southeast Asian traders.

With no large predators around other than descendants of pet dogs, cows and pigs will probably own Hawaii. Elsewhere, dogs may even help livestock survive: sheep ranchers in Tierra del Fuego often swear that the shepherding instinct is so deeply bred in their kelpie dogs that their own absence would be immaterial.

If, however, we humans do remain at the top of the planetary pecking order, in such numbers that ever more wilderness is sacrificed to food production, Peter Ward's scenario is conceivable, although total human dominion over nature won't ever happen. Small, fast-reproducing animals like rodents and snakes adapt to anything short of glaciers, and both will be continually selected for fitness by feral cats, highly fertile themselves. In his book *Future Evolution,* Ward imagines rats that evolve into kangaroo-sized hoppers with saber tusks, and snakes that learn to soar.

Frightening or entertaining, at least for now that vision is fanciful. The lesson of every extinction, says the Smithsonian's Doug Erwin, is that we can't predict what the world will be 5 million years later by looking at the survivors.

"There will be plenty of surprises. Let's face it: who would've predicted the existence of turtles? Who would ever have imagined that an organism would essentially turn itself inside out, pulling its shoulder girdle inside its ribs to form a carapace? If turtles didn't exist, no vertebrate biologist would've suggested that anything would do that: he'd have been laughed out of town. The only real prediction you can make is that life will go on. And that it will be interesting."

PART IV

CHAPTER 17

꿍

Where Do We Go from Here?

"IF HUMANS WERE gone," says ornithologist Steve Hilty, "at least a third of all birds on Earth might not even notice."

He's referring to the ones that don't stray from isolated Amazon jungle basins, or far-flung Australian thorn forests, or Indonesian cloud slopes. Whether other animals who probably *would* notice—stressed, hunted, and endangered bighorn sheep or black rhinos, for instance—would actually celebrate our passing is beyond our understanding. We can read the emotions of very few animals, most of them tame, like dogs and horses. They would miss the steady meals and, despite those leashes and reins, maybe some kindly owners. Animal species we consider the most intelligent—dolphins, elephants, pigs, parrots, and our chimpanzee and bonobo cousins—probably wouldn't miss us much at all. Although we often go to considerable lengths to protect them, the danger usually is us.

Mainly, we'd be mourned by creatures who literally can't live without us because they've evolved to live *on* us: *Pediculus humanus capitis* and her brother *Pediculus humanus humanus*—respectively, head and body lice. The latter are so specifically adapted that they depend not just on us, but on our clothing—a trait unique among species, save maybe fashion designers. Also bereaved will be follicle mites, so tiny that hundreds live even in our eyelashes, helpfully munching on skin cells as we discard them, lest dandruff overwhelm us.

Some 200 bacteria species also call us home, especially those dwelling in our large intestines and nostrils, inside our mouths, and on our teeth.

And hundreds of little *Staphylococci* live on every square inch of our skin, with thousands in our armpits and crotches and between our toes. Nearly all are so genetically customized to us that when we go, so do they. Few would attend a farewell banquet on our corpses, not even the follicle mites: Contrary to a widespread myth, hair doesn't keep growing after death. As our tissues lose moisture, they contract; the resulting exposed hair roots make exhumed cadavers look in need of a trim.

If we all suddenly keeled over en masse, the usual scavengers would clean our bones within a few months, save for anyone whose mortal husk dropped into a glacial crevasse and froze, or who landed in mud deep enough to be covered before oxygen and the biological wrecking crew started in. But what about our dear departed who preceded us into whatever comes next, whom we carefully and ritually laid to rest? How long do human remains, well, remain? Will humankind approach immortality at least as recognizably as the Barbie and Ken dolls created in someone's slick idea of our image? How long do our extensive, and expensive, efforts to preserve and seal away the dead actually last?

In much of the modern world, we begin with embalming, a gesture that delays the inevitable very temporarily, says Mike Mathews, who teaches the process in the University of Minnesota's Mortuary Science program, as well as chemistry, microbiology, and funeral history.

"Embalming's really just for funerals. The tissues coagulate a bit, but they start breaking down again." Because it's impossible to completely disinfect a body, Mathews explains, Egyptian mummifiers removed all organs, where decomposition inevitably begins.

Bacteria left in the intestinal tract are soon aided by natural enzymes that become active as a dead body's pH changes. "One of them is the same as Adolph's Meat Tenderizer. They break down our proteins so they're easier to digest. Once we stop, they kick in, embalming fluid or not."

Embalming was uncommon until the Civil War, when it was used to send fallen soldiers home. Blood, which decomposes rapidly, was replaced with anything handy that didn't. Often, it was whiskey. "A bottle of scotch works fine," allows Mathews. "It's embalmed me several times."

Arsenic turned out to work even better, and was cheaper. Until it was banned in the 1890s, it was used widely, and heavy arsenic levels are some-times a problem for archaeologists examining some old U.S. graveyards. What they generally find is that the bodies decomposed anyway, but the arsenic stayed.

After that came today's formaldehyde, from the same phenols that pro-duced Bakelite, the first man-made plastic. In recent years, a green burial movement has protested formaldehyde, which oxidizes to formic acid, the toxin in fire ants and bee stingers, as yet one more poison to leach into wa-ter tables: careless people, polluting even from the tomb. The eco-burialists also challenge why, after we intone sacred words about dust returning to dust, we ambivalently place bodies in the ground, yet go to extraordinary lengths to seal them away from it.

That sealing begins—but only begins—with the casket. Pine boxes have yielded to modern sarcophagi of bronze, pure copper, stainless steel, or coffins crafted from an estimated 60 million board feet of temperate and tropical hardwoods, felled annually just to be buried underground. Yet not really under the ground, because the box into which we're tucked for good goes inside another box, a liner usually made of plain gray con-crete. Its purpose is to support the weight of the earth so that, as in older cemeteries, graves don't sink and headstones don't tumble when caskets below rot and collapse. Because their lids aren't waterproof, holes in liner bottoms allow whatever trickles in to drain away.

The green burial folks prefer no liners, and coffins of materials that quickly biodegrade, like cardboard or wicker—or none at all: unem-balmed, shrouded bodies are placed right in the dirt to start returning their leftover nutrients to the earth. Although most people throughout history were probably interred this way, in the Western world only a handful of cemeteries permit it—and even less, the green headstone sub-stitute: planting a tree to immediately harvest the formerly human nour-ishment.

The funeral industry, emphasizing the value of preservation, counsels something rather more substantial. Even concrete liners are considered crude compared to bronze vaults so tight that in a flood, they pop up and float, despite weighing as much as an automobile.

According to Michael Pazar, vice president at Wilbert Funeral Services of Chicago, the biggest manufacturer of such burial bunkers, the challenge is that "tombs, unlike basements, don't have sump pumps." His company's triple-layered solution is pressure-tested to withstand a six-foot head of water—meaning a cemetery transformed by a rising water table into a pond. It has a concrete core, clad in rust-proof bronze, lined inside and sheathed outside with ABS: an alloy of acrylonitrile, styrene, and butadiene rubber, which may be the most indestructible, impact-and-heat-resistant plastic there is.

Its lid is affixed with a proprietary butyl sealer that bonds to the seamless plastic liner. The sealer, Pazar says, may be strongest of all. He mentions a major private testing lab in Ohio, whose reports are also proprietary. "They heated it, hit it with ultraviolet, soaked it in acid. The test report said it will last millions of years. I'd be uncomfortable with that, but these guys are Ph.D.s. Imagine some time in the future when archaeologists only find these rectangular butyl rings."

What they won't find, however, is much sign of the former human to whom we devoted all the expense, chemistry, radiation-resistant polymers, endangered hardwoods, and heavy metals—which, like the mahogany and walnut, were wrenched from the Earth only to be stuck back in it. With no incoming food to process, the body's enzymes will have liquefied whatever tissues bacteria didn't eat, mixing the results for a few decades with the acidic stew of embalming juices. That will be yet another test for the seals and ABS plastic liner, but they should easily pass, outlasting even our bones. Should those archaeologists arrive before the bronze and concrete and everything else but the butyl seal have dissolved away, all that remains of us will be a few inches of human soup.

Deserts like the Sahara, the Gobi, and Chile's Atacama, where desiccation is near-total, occasionally yield natural human mummies, with intact clothing and hair. Thawing glaciers and permafrost sometimes give up other long-dead, eerily preserved predecessors of we, the living, such as the leather-clad Bronze Age hunter discovered in 1991 in the Italian Alps.

There won't be much chance, however, for any of us presently alive to leave a lasting mark. It's rare these days for anyone to be covered with mineral-rich silt that eventually replaces our bone tissue until we turn into skeleton-shaped rocks. In one of our stranger follies, we deny ourselves and our loved ones the opportunity of a true lasting memorial—

fossilhood—with extravagant protections that, in the end, only protect the Earth from being tainted by us.

⁓

THE ODDS OF us all going together, let alone soon, are slight, but within the realm of possibility. The chance that only humans will die, leaving everything else to carry on, is even more remote, but nevertheless greater than zero. Dr. Thomas Ksiazek, chief of the Special Pathogens Branch at the U.S. Centers for Disease Control, is paid to worry that something could take out many millions of us. Ksiazek is a former army veterinary microbiologist and a virologist, and his consultations range from threats of biological attack to hazards that unexpectedly jump from other species, such as the SARS coronavirus he helped to characterize.

Grim as those scenarios are, especially in an age when so many of us live in oversized Petri dishes called cities, where microbes congregate and flourish, he doesn't see an infectious agent arising that could wipe out the entire species. "It would be unparalleled. We work with the most virulent, and even with those there are survivors."

In Africa, periodic horrors like the Ebola and Marburg viruses have slain villagers, missionaries, and so many health care workers that the rest fled their hospitals. In each instance, what finally broke the chain of contagion was simply getting staff to wear protection and scrub with soap and water—often lacking in poor areas where such diseases usually begin—after touching patients.

"Hygiene is the key. Even if someone tried to introduce Ebola intentionally, though you might get a few secondary cases in families and hospital staff, with sufficient precautions it would die out rapidly. Unless it mutated to something more viable."

High-hazard viruses like Ebola and Marburg originate in animals—fruit bats are suspected—and are spread among people through infected body fluids. Since Ebola finds its way into the respiratory tract, U.S. Army researchers at Fort Detrick, Maryland, tried to see if a terrorist might be able to concoct an Ebola bomb. They created an aerosol capable of spreading the virus back to animals. "But," says Ksaizek, "it doesn't make respiratory particles small enough to be easily transmissible to humans via coughing or wheezing."

But if one Ebola strain, *Reston,* ever mutated, we might have a problem. Currently, it kills only nonhuman primates; unlike other Ebolas, however, it is believed to attack through the air. Similarly, if highly virulent AIDS, which is currently passed through blood or semen, were ever to become airborne, it could be a real species-stopper. That's unlikely, Ksiazek believes.

"Possibly it could change its transmission route. But the current way is actually advantageous to HIV's survival because it allows victims to spread it around awhile. It evolved into that niche for a reason."

Even the deadliest airborne influenzas have failed to wipe out everyone, because people eventually develop immunity and pandemics fizzle. But what if a psychotically obsessed, biochemically trained terrorist creatively spliced something together that evolves faster than we develop resistance—maybe by clipping genetic material into the versatile SARS virus, which could spread both sexually and via the air before Ksaizek helped eradicate it?

It would be possible to design for extreme virulence, Ksaizek allows, although, as in transgenic pesticides, results of genetic manipulation aren't guaranteed.

"It's like when they breed mosquitoes to be less capable of transmitting a viral disease. When they release these lab-bred mosquitoes, they don't compete very well. It's not as easy as just thinking about it. Synthesizing a virus in the lab is one thing; making it work is another. In order to repackage it as an infectious virus, you need a constellation of genes that will let it infect a host cell, then make a bunch of progeny."

He chuckles mirthlessly. "People trying might kill themselves in the process. There are a lot of easier things to do with a lot less effort."

HAVING YET TO perfect contraception, thus far we have little to fear from misanthropic plots to sterilize the entire human race. From time to time, Nick Bostrom, who directs Oxford's Future of Humanity Institute, computes the odds (increasing, he believes) that human existence is at risk of ending. He is particularly intrigued by the potential of nanotechnology going awry, accidentally or deliberately, or superintelligence running amuck. In either case, however, he notes that the skills needed to create atom-sized medical machines that would patrol our bloodstreams, zapping disease until they suddenly turned on us, or self-replicating robots that end up crowding or outsmarting us off the planet, are "at least decades away."

In his morose 1996 scholarly tome *The End of the World,* cosmologist John Leslie of Ontario's University of Guelph concurs with Bostrom. He cautions, however, that there's no assurance that our current dallying with high-energy particle accelerators won't crack the very physics of the vacuum in which our galaxy twirls, or even touch off a whole new Big Bang ("by mistake," he adds, with scant consolation).

Each of these men, philosophers taking ethical measure of an age in which machines think faster than humans but regularly prove at least as flawed, repeatedly smack into a phenomenon that never troubled their intellectual predecessors: although humans have obviously survived every pox and meteor that nature has tossed at us until now, technology is something we toss back at our own peril.

"On the bright side, it hasn't killed us yet, either," says Nick Bostrom, who, when not refining doomsday data, researches how to extend the human life span. "But if we did go extinct, I think it would more likely be through new technologies than environmental destruction."

To the rest of the planet, it would make little difference, because if either actually took us out, many other species undoubtedly would go with us. The chance that zookeepers from outer space might make this whole conundrum moot by rapturing us away but leaving everything else is not only slim but narcissistic—why would they only be interested in us? And what would stop them from the alien equivalent of salivating over the same enticing resource repast that we've gorged on? Our seas, forests, and the creatures that dwell in them might quickly prefer us to hyper-powered extraterrestrials who could stick an interstellar straw into the planetary ocean for the same purposes that induce us to siphon entire rivers out of their valleys.

"By definition, we're the alien invader. Everywhere except Africa. Every time *Homo sapiens* went anywhere else, things went extinct."

Les Knight, the founder of VHEMT—the Voluntary Human Extinction Movement—is thoughtful, soft-spoken, articulate, and quite serious.

Unlike more-strident proponents of human expulsion from an aggrieved planet—such as the Church of Euthanasia, with its four pillars of abortion, suicide, sodomy, and cannibalism, and a website guide to butchering a human carcass that includes a recipe for barbeque sauce—Knight takes no misanthropic joy in anyone's war, illness, or suffering. A schoolteacher, he just keeps doing math problems that keep giving him the same answer.

"No virus could ever get all 6 billion of us. A 99.99 percent die-off would still leave 650,000 naturally immune survivors. Epidemics actually strengthen a species. In 50,000 years we could easily be right back where we are now."

War doesn't work either, he says. "Millions have died in wars, and yet the human family continues to increase. Most of the time, wars encourage both winners and losers to repopulate. The net result is usually an increase rather than a decrease in total population. Besides," he adds, "killing is immoral. Mass murder should never be considered a way to improve life on Earth."

Although he lives in Oregon, his movement, he says, is based everywhere—meaning on the Internet, with websites in 11 languages. At Earth Day fairs and environmental conferences, Knight posts charts that acknowledge U.N. predictions that, worldwide, population growth rates and birthrates will both decline by 2050—but the punch line is the third chart, which shows sheer numbers still soaring.

"We have too many active breeders. China's down to 1.3 percent reproduction, but still adds 10 million a year. Famine, disease, and war are harvesting as fast as ever, but can't keep up with our growth."

Under the motto "May we live long and die out," his movement advocates that humanity avoid the agonizing, massive die-off that will occur when, as Knight foresees, it becomes brutally clear that it was naive to think that we could all have our planet and eat it, too. Rather than face horrific resource wars and starvation that decimate us and nearly everything else as well, VHEMT proposes gently laying the human race to rest.

"Suppose we all agree to stop procreating. Or that the one virus that would truly be effective strikes, and all human sperm loses viability. The first to notice would be crisis-pregnancy centers, because no one would be coming in. Happily, in a few months abortion providers would be out of business. It would be tragic for people who kept trying to conceive. But in five years, there would be no more children under five dying horribly."

The lot of all living children would improve, he says, as they became more valuable rather than more disposable. No orphan would go un-adopted.

"In 21 years, there would be, by definition, no juvenile delinquency." By then, as resignation sinks in, Knight predicts that spiritual awakening would replace panic, because of a dawning realization that as human life drew toward a close, it was improving. There would be more than enough to eat, and resources would again be plentiful, including water. The seas would replenish. Because new housing wouldn't be necessary, so would forests and wetlands.

"With no more resource conflicts, I doubt we'd be wasting each other's lives in combat." Like retired business executives who suddenly find serenity by tending a garden, Knight envisions us spending our remaining time helping rid an increasingly natural world of unsightly and now use-less clutter, in pursuit of which we'd once swapped something alive and lovely.

"The last humans could enjoy their final sunsets peacefully, knowing they have returned the planet as close as possible to the Garden of Eden."

⁘

IN AN AGE when the decline of natural reality is paralleled by the rise of something called virtual reality, VHEMT's antipode is not just those who find the promise of better living through human extinction deranged, but also a group of respected thinkers and noted inventors who consider ex-tinction possibly a career move for *Homo sapiens.* Transhumanists, as they call themselves, hope to colonize virtual space by developing software to upload their minds into circuitry that would outperform both our brains and bodies on numerous levels (including, incidentally, never having to die). Via the self-accruing wizardry of computers, an abundance of sili-con, and vast opportunities afforded by modular memory and mechanical appendages, human extinction would become merely a jettisoning of the limited and not very durable vessels that our technological minds have finally outgrown.

Prominent in the transhumanist (sometimes called posthuman) move-ment are Oxford philosopher Nick Bostrom; heralded inventor Ray Kurzweil, originator of optical character recognition, flat-bed scanners, and print-to-speech reading machines for the blind; and Trinity College bioethicist James Hughes, author of *Citizen Cyborg: Why Democratic Societies Must Respond to the Redesigned Human of the Future*. However Faustian, their discussion is compelling in its lure of immortality and preternatural power—and almost touching in its utopian faith that a machine could be made so perfect that it would transcend entropy.

The great barrier to robots and computers leaping the chasm between mere objects and life-forms, it's often argued, is that no one has ever built a machine that is aware of itself: without being able to feel, a supercomputer might calculate rings around us, but still never be able to think about its place in the world. A more fundamental flaw, though, is that no machine has performed indefinitely without human maintenance. Even stuff without moving parts breaks down, and self-repair programs crash. Salvation, in the form of backup copies, could lead to a world of robots desperately trying to stay one clone up on the latest technology to which competitive knowledge was migrating—an all-consuming form of tail chasing that hearkens to the behavior of lower primates, who undoubtedly have more fun.

Even if posthumanists succeed in transferring themselves to circuitry, it won't be anytime soon. For the rest of us, sentimentally clinging to our carbon-based human nature, voluntary-extinction advocate Les Knight's twilight prophecy hits a vulnerable spot: the weariness that genuinely humane beings feel as they witness the collapse of much biology and beauty. The vision of a world relieved of our burden, with its flora and fauna blossoming wildly and wonderfully in every direction, is initially seductive. Yet it's quickly followed by a stab of bereavement over the loss of all the wonder that humans have wrought amid our harm and excess. If that most wondrous of all human creations—a child—is never more to roll and play on the green Earth, then what really would be left of us? What of our spirit might be truly immortal?

Deferring for the moment the matter of afterlife as defined by religions great and small: After we're all gone, what would become of the passion that believers and agnostics alike share—our irrepressible need to utter what's in our souls? What will remain of our greatest creative forms of human expression?

CHAPTER 18

Art Beyond Us

BEHIND A CONVERTED Tucson warehouse that houses the Metal-physic Sculpture Studio, two foundry workers don rough-hide jackets and chaps, gloves of asbestos and stainless-steel mesh, and hard hats with eye shields. From a firebrick kiln, they remove preheated ceramic molds of the sculpted wings and body of an African white-backed vulture, which, once cast and welded together, will form a life-size bronze for the Philadelphia Zoo by wildlife artist Mark Rossi. They position these, sprue channels pointed upward, in a sand-filled turntable that slides on a track over to a drum-shaped, steel-clad liquid-propane furnace. The 20-pound ingots they loaded inside earlier have decomposed into 2,000°F bronze soup, sloshing on the same heat-resistant ceramic used for space shuttle tiles.

The furnace is tilt-mounted on an axle, so little effort is required to pour molten metal into the waiting molds. Six thousand years ago in Persia, the fuel was cordwood, and the molds were cavities in clay hillsides, not ceramic shells. But except for the copper-silicon alloy favored today over the copper-arsenic or copper-tin blends the ancients used, the process of immortalizing art in bronze is essentially the same.

And for the same reasons: Copper, like silver and gold, is one of the noble metals, resistant to corrosion. Some of our ancestors first noticed it oozing like honey from a piece of malachite near a campfire. When it cooled, they found it malleable, durable, and quite beautiful. They tried melting other rocks, mixed the results, and man-made metal alloys of unprecedented strength were born.

Some of the rocks they tested contained iron, a tough base metal, but one that oxidized rapidly. It proved more resistant when mixed with carbon ash, and even stronger after laborious hours of pumping a bellows to blast out excess carbon. The result was just enough forged steel for a few prized Damascus swords but not much else until 1855, when Henry Bessemer's high-powered blowers finally turned steel from a luxury into a commodity.

But don't be fooled, says David Olson, head materials scientist at the Colorado School of Mines, by massive steel buildings, steamrollers, tanks, railway tracks, or the shine on your stainless cutlery. Bronze sculpture will outlast all of it.

"Anything made of noble metals likely will exist forever. Any metal that comes from a mineral compound like iron oxide will go back to that compound. It was there for millions of years. We've just borrowed it from the oxygen and pumped it to a higher energy state. It all falls back there."

Even stainless steel: "It's one of many fantastic alloys designed to perform a specific service. In your kitchen drawer, it stays beautiful forever. Leave it in oxygen and salt water, it's on its way out."

Bronze artwork is doubly blessed. Scarce, expensive noble metals, like gold, platinum, and palladium, combine with nearly nothing in nature. Copper, more plentiful and slightly less regal, forms bonds when exposed to oxygen and sulfur, but—unlike iron, which crumbles as it rusts—the result is a film, two-thousandths to three-thousandths of an inch thick, that protects it from further corruption. These patinas, lovely in their own right, form part of the allure of bronze sculpture, which is at least 90 percent copper. Besides adding strength and making copper easier to weld, alloys can simply make it harder. One icon of Western culture that Olson expects to last long will be pre-1982 copper pennies (actually, they're bronze, containing 5 percent zinc). Today, however, the U.S. cent is nearly all zinc, with only enough copper to memorialize the color of money once worth its face value.

That new, 97.6 percent zinc penny will leach away if tossed in the ocean, dooming Abe Lincoln's visage to be filtered by shellfish within a century or so. The Statue of Liberty, however, which sculptor Frederic Auguste Bartholdi hammered from copper sheeting not much thicker, would oxidize with dignity at the bottom of New York Harbor should glaciers ever return to our warming world and knock her off her pedestal. In the end, Liberty's sea green patina will thicken until she turns to stone, but the sculptor's aesthetic intention will still be preserved for the fish to ponder. By then, Africa's white-backed vultures may also have vanished, except in Mark Rossi's bronze homage to them, in whatever is left of Philadelphia.

Even if the primeval Białowieża Puszcza forest spreads anew across Europe, the bronze memorial to its founder, horseback King Jagiełło in New York's Central Park, will probably outlast it one distant day when the aging sun overheats and life on Earth finally winds down. In their Central Park West studio northwest of his statue, Manhattan art conservators Barbara Appelbaum and Paul Himmelstein coax fine old materials to remain in the high-energy state to which artists have taken them. They are acutely aware of the lasting power of things elemental.

"What we know of ancient textiles in China," says Himmelstein, "is because silk was used to wrap bronzes." Long after it disintegrated, the fabric's texture remained imprinted in the copper salts of the patina. "And all we know of Greek textile is from paintings on fired ceramic vases."

Ceramics, being minerals, are as close to their lowest energy state as things get, says Appelbaum, who has high-energy dark eyes framed by short-cropped white hair. From a shelf she produces a baby trilobite, mineralized in faithful detail by Permian mud, exquisitely readable 260 million years later. "Unless you smash them, ceramics are virtually indestructible."

Unfortunately, that happens, and tragically, most of history's bronze statues are also gone, melted down for weapons. "Ninety-five percent of all artwork ever made doesn't exist anymore," says Himmelstein, a knuckle stroking his gray goatee. "We know little of Greek or Roman painting—mainly just what writers like Pliny tell us about it."

On a masonite table lies a large oil they are repairing for a private collector, a 1920s portrait of a moustachioed Austro-Hungarian noble with a jeweled watch fob. It had sagged and begun to molder after years in some dank hallway. "Unless they're hanging in 4,000-year-old pyramids with zero moisture, within a few hundred years of neglect, paintings on canvas will be a dead issue."

Water, the stuff of life, is often the death of art—unless the art is submerged in it.

"If space aliens show up after we're gone and all the museum roofs have leaked and everything inside has rotted, they should dig up the deserts and dive underwater," Himmelstein says. If the pH isn't too acidic, the lack of oxygen can even preserve waterlogged textiles. Removing them from the water can be perilous—even copper that lies for millennia in chemical equilibrium with seawater may develop "bronze disease" outside of it, due to reactions that turn chlorides into hydrochloric acid.

"On the other hand," says Appelbaum, "we tell people who ask advice about time capsules that good-quality rag paper in an acid-free box should last forever, as long as it never gets wet. Just like Egyptian papyrus." Immense archives of acid-free paper, including the world's largest collection of photographs, owned by the stock photo agency Corbis, have been climatically sealed in a former limestone mine in western Pennsylvania, 200 feet below ground. The vault's dehumidifiers and subzero refrigeration are guaranteed to secure them for at least 5,000 years.

Unless, of course, the power goes off. Despite our best efforts, things do go amiss. "Even in dry Egypt," notes Himmelstein, "the most valuable library yet assembled—a half-million papyrus scrolls in Alexandria, some of them Aristotle's—was perfectly preserved until a bishop lit a torch to expel paganism."

He wipes his hands on his blue pinstriped apron. "At least we know about them. The saddest thing is that we have no idea of what ancient music was like. We have some of the instruments. But not the sounds made on them."

Neither of these esteemed conservators figures that music as it is recorded today—nor any other information stored on digital media—has much chance to survive, let alone be apprehended by any sentient being that might puzzle over a stack of flimsy plastic disks in the distant future. Some museums now use lasers to etch knowledge microscopically on stable copper—a good idea, assuming the mechanisms to read them survive with them.

And yet, of all human creative expression, it happens that music may have the best chance of all to echo on.

❧

In 1977, Carl Sagan asked Toronto painter and radio producer Jon Lomberg how an artist might express the essence of human identity to an audience that had never seen humans. With fellow Cornell astrophysicist Frank Drake, Sagan had just been invited by NASA to devise something meaningful about humanity to accompany the twin Voyager spacecrafts, which would visit the outer planets and then continue on through interstellar space, possibly forever.

Sagan and Drake had also been involved with the only other two space probes to leave the solar system behind. *Pioneer 10* and *Pioneer 11* were launched in 1972 and 1973, respectively, to see if the asteroid belt could be navigated and to inspect Jupiter and Saturn. *Pioneer 10* survived a hot 1973 encounter with radioactive ions in Jupiter's magnetic field, sent back images of Jovian moons, and kept going. Its last audible transmission was in 2003; at the time, it was nearly 8 billion miles from Earth. In 2 million years, it should pass, but not dangerously near, the red star Aldebran, the eye in the constellation Taurus. *Pioneer 11* whipped around Jupiter a year after its sibling, using its gravity like a sling to propel it past Saturn in 1979. Its escape trajectory sent it in the direction of Sagittarius; it won't pass any stars for 4 million years.

Both Pioneers carry 6-by-9-inch gold-plated aluminum plaques bolted to their frames, bearing line etchings by Sagan's former wife Linda Salzman that depict a naked human male and female. Next to them are graphical depictions of Earth's position in the solar system and the sun's location in the Milky Way, plus the cosmic equivalent of a phone number: a mathematical key based on a transitional state of hydrogen, indicating wavelengths where we're tuned in, listening.

The messages carried by the Voyagers, Sagan told Jon Lomberg, would go into much more detail about us. In an era preceding digital media, Drake had contrived a way to record both sounds and images on a 12-inch, gold-plated copper analog disk, which would include a stylus and, they hoped, intelligible diagrams on how to play it. Sagan wanted Lomberg, the illustrator of his popular books, as the recording's design director.

The notion was boggling: conceive and choreograph a showcase that would be a work of art in itself, bearing what might likely be the last remaining fragments of human aesthetic expression. Once aloft, the gold-anodized aluminum box containing the record, whose cover Lomberg would also design, would be exposed to weathering by cosmic rays and interstellar dust. By conservative estimates, it would last at least a billion years, but probably much longer. By then, tectonic upheavals or an expanded sun might well have rendered any signs of us left on Earth down to their molecular essence. It might be the closest that any human artifact would get to a chance at eternity.

Lomberg had only six weeks to think about that before launch. He and his colleagues polled world figures, semioticians, thinkers, artists, scientists, and science-fiction writers on what might possibly penetrate the consciousness of unfathomable viewers and listeners. (Years later, Lomberg would also help design the warning to trespassers of buried radioactive peril at New Mexico's Waste Isolation Pilot Plant.) The disk would carry recorded greetings in 54 human languages, plus voices of dozens of other Earth inhabitants, from sparrows to whales, and sounds such as a heartbeat, surf, a jackhammer, crackling fire, thunder, and a mother's kiss.

The pictures included diagrams of DNA and the solar system, as well as photographs of nature, architecture, town and cityscapes, women nursing babies, men hunting, children contemplating a globe, athletes competing, and people eating. Since the finders might not realize that a photo was more than abstract squiggles, Lomberg sketched some accompanying silhouettes to help them discern a figure from its background. For a portrait of a five-generation family, he silhouetted individuals and included notations conveying their relative sizes, weights, and ages. For a human couple, he made the woman's silhouetted womb transparent to reveal the fetus growing within, hoping that communion between an artist's idea and an unseen viewer's imagination might transcend even enormous time and space.

"My job was not just to find all these images, but to sequence them in a way that added more information than the sum of the individual pictures," he recalls today in his home near Hawaii's observatory-studded Mauna Kea volcano. Beginning with things a cosmic traveler might recognize, such as planets as seen from space or the spectra of stars, he arranged images along an evolutionary flow, from geology to the living biosphere to human culture.

Similarly, he orchestrated the sounds. Although he was a painter, he sensed that music had a better chance than images to reach, and maybe even enchant, the alien mind. Partly, because rhythm is manifest throughout physics, but also because for him, "other than nature, it's the most reliable way to get into touch with what we call spirit."

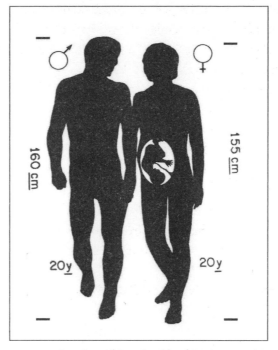

Diagram of male and female, drawn by Jon Lomberg
for the *Voyager* Spacecraft Golden Record.

ARTWORK BY JON LOMBERG/© 2000.

The disk contains 26 selections, including music of pygmies, Navajos, Azerbaijani bagpipes, mariachis, Chuck Berry, Bach, and Louis Armstrong. Lomberg's most cherished nominee was the Queen of the Night's aria from Mozart's *The Magic Flute*. In it, soprano Edda Moser, backed by the Bavarian State Opera Orchestra, displays the upper limit of the human voice, hitting the loftiest note in the standard operatic repertoire, a high F. Lomberg and the record's producer, former *Rolling Stone* editor Timothy Ferris, insisted to Sagan and Frank Drake that it be included.

They quoted Kierkegaard, who had once written: "Mozart enters that small, immortal band whose names, whose works, time will not forget, for they are remembered in eternity."

With Voyager, they felt honored to make that truer than ever.

The two Voyagers were launched in 1977. Both passed Jupiter in 1979 and reached Saturn two years later. After its sensational discovery of active volcanoes on Jupiter's moon Io, *Voyager 1* dipped below Saturn's south pole for our first glimpse of its moon Titan, which flipped it out of the solar system's elliptical plane and off toward interstellar space, actually passing *Pioneer 10*. It is now farther from Earth than any other human-made object. *Voyager 2* took advantage of a rare planetary alignment to visit Uranus and Neptune, and now is also leaving the sun behind.

Lomberg watched the first Voyager launch, with the record's gilded sleeve bearing his diagrams of its birthplace and what to do with the disk inside—glyphs that he, Sagan, and Drake hoped that any space-navigating intelligence would be able to decipher, though there was little chance it would ever be found, and even less that we would ever know about it. Yet neither the Voyagers nor their recordings are the first man-made entities to travel beyond our planetary neighborhoods. Even after billions of years of relentless space-dust abrasion wears them to dust themselves, there is yet another chance for us to be known beyond our world.

DURING THE 1890s, a Serbian immigrant to America, Nikola Tesla, and an Italian, Guglielmo Marconi, each patented devices capable of sending wireless signals. In 1897, Tesla demonstrated sending ship-to-shore pulses across bodies of water in New York, even as Marconi was doing the same among various British isles—and, in 1901, across the Atlantic. Eventually they sued each other over the claim, and the royalties, to the invention of radio. No matter who was right, by then transmission across seas and continents was routine.

And beyond: Electromagnetic radio waves—waves much longer than poisonous gamma radiation or ultraviolet sunlight—emanate at the speed of light in an expanding sphere. As they move outward, their intensity drops by a factor of one over the distance squared, meaning that at 100 million miles from Earth, the signal strength is one-fourth what it was at 50 million miles. Nevertheless, it is still there. As the sphere of a transmission's surface expands through the Milky Way, galactic dust absorbs some of the radio radiation, attenuating the signal further. Still, it keeps going.

In 1974, Frank Drake beamed a three-minute radio greeting from the largest radio dish on Earth, the 1,000-foot, half-million-watt Arecibo Radio Telescope in Puerto Rico. The message consisted of a series of binary pulses that an extraterrestrial mathematician might recognize as representing a crude graphical arrangement, depicting the sequence 1 through 10, the hydrogen atom, DNA, our solar system, and a human-shaped stick figure.

The signal, Drake later explained, was about a million times stronger than a typical TV transmission, and was aimed at a star cluster in the constellation Hercules, where it wouldn't arrive for 22,800 years. Even so, due to the subsequent outcry over possibly having revealed Earth's whereabouts to superior, predatory alien intelligences, members of the international community of radio astronomers agreed to never unilaterally expose the planet to such a risk again. In 2002, that accord was ignored by Canadian scientists who directed lasers heavenward. But as Drake's broadcast has yet to elicit a response, let alone an attack, the chance that anything might cross their tight beams can't be meaningfully computed.

Besides, the cat may long be out of the bag. For more than a half-century, we've been sending signals that by now would take a very large or very sensitive receiver to collect—yet, considering the size of the intellect that we imagine might be out there, it's not impossible.

In 1955, a little more than four years after leaving a TV studio in Hollywood, signals bearing the first sound and images of the *I Love Lucy* show passed Proxima Centauri, the nearest star to our sun. A half-century later, a scene with Lucy disguised as a clown sneaking into Ricky's Tropicana Night Club was 50-plus light-years, or about 300 trillion miles, away. Since the Milky Way is 100,000 light-years across and 1,000 light-years thick, and our solar system is near the middle of the galactic plane, this means in about AD 2450 the expanding sphere of radio waves bearing Lucy, Ricky, and their neighbors the Mertzes will emerge from the top and bottom of our galaxy and enter intergalactic space.

Before them will lie billions of other galaxies, over distances we can quantify but can't really comprehend. By the time *I Love Lucy* reaches them, it's unclear how anything out there would be able to make much sense of it, either. Distant galaxies, from our perspective, are moving away from each other, and the farther away they are, the faster they move—an astronomical quirk that appears to define the very fabric of space itself. The farther radio waves go, the weaker they become, and the longer they appear. Out at the universe's edge, 10 billion-plus light-years away from now, light from our galaxy seen by some superintelligent race would appear shifted to the red end of the spectrum, where the longest wavelengths lie.

Massive galaxies in their path would further distort radio waves bearing the news that in 1953, a baby boy was born to Lucille Ball and Desi Arnaz. It would also increasingly compete with the background noise from the Big Bang, the original birth cry of the universe, which a consensus of scientists dates to at least 13.7 billion years ago. Just like Lucy's broadcast shenanigans, that sound has been expanding at the speed of light ever since, and thus pervades everything. At some point, radio signals become even weaker than that cosmic background static.

But however fragmented, Lucy would be there, even fortified by the far more robust ultrahigh-frequency broadcasts of her reruns. And Marconi and Tesla, the most gossamer of electronic ghosts by now, would have preceded her, and Frank Drake after them. Radio waves, like light, keep expanding. To the limits of our universe and our knowledge, they are immortal, and broadcast images of our world and our times and memory are there with them.

As the Voyagers and Pioneers erode away to stardust, in the end our radio waves, bearing sounds and images that record barely more than a single century of human existence, will be all the universe holds of us. It's hardly an instant, even in human terms, but a remarkably fruitful—if convulsive— one. Whoever awaits our news at the edge of time will get an earful. They may not understand *Lucy,* but they will hear us laugh.

❧

CHAPTER 19

The Sea Cradle

THE SHARKS HAVE never seen humans before. And few humans present have ever seen so many sharks.

Except for moonlight, the sharks have also never seen the equatorial night be anything other than dark and deep. Nor have the eel fish, which resemble 5-foot silver ribbons with fins and needle snouts as they skitter up to the research vessel *White Holly*'s steel hull, entranced by shafts of color drilled into the night sea by spotlights from the captain's deck. Too late, they notice that the waters here are boiling with dozens of white-tipped, black-tipped, and gray reef sharks racing in delirious circles that scream hunger.

A quick squall comes and goes, blowing a curtain of warm rain across the lagoon where the ship is anchored and drenching the remains of a deckside chicken dinner eaten over a plastic tarp stretched across the dive master's table. Still, the scientists linger at the *White Holly*'s railing, fascinated by thousands of pounds of sharks—sharks proving that they rule the food pyramid here by snatching eel fish in mid-flight as they leap between swells. Twice a day for the past four days, these people have swum among such sleek predators, counting them and everything else alive in the water, from swirling rainbows of reef fish to iridescent coral forests; from giant clams lined with velvety, multihued algae down to microbes and viruses.

This is Kingman Reef, one of the hardest places to reach on Earth. To the naked eye, it barely exists: a change from cobalt blue to aquamarine is

Gray reef shark. *Carcharhinus amblyrhynchos.* Kingman Reef.
J. E. MARAGOS, U.S. FISH AND WILDLIFE SERVICE.

the main clue that a nine-mile-long coral boomerang lies submerged 15 meters below the surface of the Pacific, 1,000 miles southwest of Oahu. At low tide two islets rise barely a meter above the water, mere slivers consisting of giant clamshell rubble heaped by storms against the reef. During World War II, the U.S. Army designated Kingman a way-station anchorage between Hawaii and Samoa, but never used it.

Two dozen scientists aboard the *White Holly* and their sponsor, Scripps Institution of Oceanography, have come to this water-world-without-people to glimpse what a coral reef looked like before human beings appeared on Earth. Without such a baseline, there can be little agreement on what constitutes a healthy reef, let alone on how to help nurse these aquatic equivalents of rain forest diversity back to whatever that might be. Although months of sifting data lie ahead, already these researchers have found evidence that contradicts convention, and seems counterintuitive even to themselves. But there it is, thrashing just off starboard.

Between these sharks and an omnipresent species of 25-pound red snappers equipped with noticeable fangs—one of which sampled a

Two-spot red snapper. *Lutjanus bohar*, Palmyra Atoll.

J. E. MARAGOS, U.S. FISH AND WILDLIFE SERVICE.

photographer's ear—it appears that big carnivores account for more total biomass than anything else here. If so, that would mean that at Kingman Reef, the conventional notion of a food pyramid is standing on its pointy head.

As ecologist Paul Colinvaux described in a seminal 1978 book, *Why Big Fierce Animals Are Rare,* most animals feed on creatures smaller and many times more numerous than themselves. Because roughly only 10 percent of the energy they consume converts to body mass, millions of little insects must feast on 10 times their total weight in tiny mites. The bugs themselves are gobbled by a correspondingly smaller number of small birds, which in turn are hunted by far fewer foxes, wildcats, and large raptors.

Even more than by head counts, Colinvaux wrote, the food pyramid's shape is defined by mass: "All the insects in a woodlot weigh many times as much as all the birds; and all the songbirds, squirrels, and mice combined weigh vastly more than all the foxes, hawks, and owls combined."

None of the scientists on this August, 2005 expedition, who hail from America, Europe, Asia, Africa, and Australia, would dispute those conclusions—on land. Yet the sea may be special. Or perhaps it's terra firma that is the exception. In a world with or without people, two-thirds of its surface is that mutable one on which the *White Holly* lightly bobs to pulsations that rock the planet. From the vantage of Kingman Reef, there are no easy contours to define our spaces, because the Pacific has no boundaries. It stretches until it blends into the Indian and the Antarctic, and squeezes through the Bering Strait into the Arctic, all of which in turn mix into the Atlantic. At one time, the Earth's great sea was the origin of everything that breathes and reproduces. As it goes, so goes everything's future.

"Slime."

Jeremy Jackson has to duck to take shade under an awning on the *White Holly's* upper deck, where the stern of this former naval cargo hauler has been converted to an invertebrate laboratory. Jackson, a Scripps marine paleoecologist with limbs and ponytail so long he suggests a king crab that short-circuited evolution, springing straight from the sea into human form, had the original idea for this mission. Jackson has spent much of his career in the Caribbean, watching the pressures of fishing and planetary warming flatten the Gruyère-cheese architecture of living coral reefs to bleached marine slag. As corals die and collapse, they and the myriad life-forms that call their crevices home, and everything that eats them, get displaced by something slick and unpleasant. Jackson leans over trays of algae that seaweed expert Jennifer Smith collected in previous stops on the way to Kingman.

"That's what we get on the slippery slope to slime," he tells her again. "Plus jellyfish and bacteria—the marine equivalent of rats and roaches."

Four years earlier, Jeremy Jackson had been invited to Palmyra Atoll, the northernmost of the Line Islands: a tiny Pacific archipelago divided by the equator and split between two nations, Kiribati and the United States. Palmyra had recently been purchased by The Nature Conservancy for coral reef research. During World War II, the U.S. Navy built an air station on Palmyra, opened channels into one lagoon, and dumped enough munitions and 55-gallon diesel drums in another that it was later dubbed Black Lagoon for its resident pool of dioxins. Except for a small U.S. Fish and Wildlife maintenance staff, Palmyra is uninhabited, its abandoned naval buildings half-dissolved into the surf. One semi-submerged boat hull is now a planter box stuffed with coconut palms. Coconut, introduced here, has all but vanquished native pisonia forests, and rats have replaced land crabs as the top predator.

Jackson's impression radically changed, however, when he jumped into the water. "I could barely see 10 percent of the bottom," he told his Scripps colleague Enric Sala when he returned. "My view was blocked by sharks and big fish. You have to go there."

Sala, a young conservation marine biologist from Barcelona, had never known big sea species in his native Mediterranean. In a strictly policed reserve off Cuba, he had seen a remnant population of 300-pound groupers. Jeremy Jackson had traced Spanish maritime records back to Columbus to verify that 800-pound versions of these monsters once spawned in huge numbers around Caribbean reefs, in the company of 1,000-pound sea turtles. On Columbus's second voyage to the New World, the seas off the Greater Antilles were so thick with green turtles that his galleons practically ran aground on them.

Jackson and Sala coauthored papers describing how our era's perspective had deluded us into thinking that a coral reef populated by colorful but puny, aquarium-sized fish was pristine. Only two centuries earlier, it was a world where ships collided with whole schools of whales, and where sharks were so big and abundant they swam up rivers to prey on cattle. The northern Line Islands, they decided, presented an opportunity to follow a gradient of decreasing human population and, they suspected, increasing animal size. At the end closest to the equator was Kiritimati, also known as Christmas Island, the world's largest coral atoll, with 10,000 people on just over 200 square miles. Next came Tabuaeran (Fanning) and a 3-square-mile speck called Teraina (Washington), with 1,900 and 900 people, respectively. Then Palmyra, with 10 staff researchers—and 30 miles farther, a sunken island where only the fringe reef that once encircled it remained: Kingman.

Other than copra—dried coconut—and a few pigs for local consumption, there is no agriculture on Kiritimati–Christmas Island. Still, in the first days of the 2005 expedition that Sala eventually organized, researchers aboard the *White Holly* were startled by the gush of nutrients from the island's four villages, and by the slime they found coating the reefs where grazers like parrotfish had been heavily fished. At Tabuaeran, rotting iron from a sunken freighter was feeding even more algae. Tiny Teraina, far overpopulated for its size, had no sharks or snappers at all. Humans there used rifles to fish the surf for sea turtles, yellowfin tuna, red-footed boobies, and melonhead whales. The reef bore a four-inch-thick mat of green seaweed.

Submerged Kingman Reef, northernmost of all, had once been the size of Hawaii's Big Island, with a volcano to match. Its caldera now lies below its lagoon, leaving only its coral ring barely visible. Because corals live in symbiosis with friendly, one-celled algae that require sunlight, as Kingman's cone keeps sinking, the reef will go, too—already its west side has drowned, leaving the boomerang shape that allowed the *White Holly* to enter and anchor in the lagoon.

"So ironic," marveled Jackson, after 70 sharks greeted the team's first dive, "that the oldest island, sinking beneath the waves like a 93-year-old man with three months before he dies, is the healthiest against the ravages of man."

Armed with measuring tape, waterproof clipboards, and three-foot PVC lances to discourage toothy natives, the teams of wet-suited scientists counted corals, fish, and invertebrates all around Kingman's broken ring, sampling up to four meters on either side of multiple 25-meter transect lines they lay beneath the transparent Pacific. To examine the microbial base of the entire reef community, they suctioned coral mucus, plucked seaweed, and filled hundreds of liter flasks with water samples.

Besides the mostly curious sharks, unfriendly snappers, furtive moray eels, and intermittent schools of five-foot barracuda, the researchers also swam through swirling shoals of fusiliers, lurking peacock groupers, hawkfish, damselfish, parrotfish, surgeonfish, befuddling variations on the yellow-blue theme of angelfish, and striped, crosshatched, and herringbone permutations of black-yellow-silver butterfly fish. The huge diversity and myriad niches of a coral reef enable each species, so close in body shape and plan, to find different ways to make a living. Some feed only on one coral, some only on another; some switch between coral and invertebrates; some have long bills to poke into interstitial spaces that conceal tiny mollusks. Some prowl the reef by daylight while others sleep, with the whole assemblage changing places at night.

"It's kind of like hot-bunking in submarines," explains Alan Friedlander of Hawaii's Oceanic Institute, one of the expedition's fish experts. "Guys take four-to-six-hour shifts, switching bunks. The bunk never stays cold for very long."

Vibrant as it is, Kingman Reef is still the aquatic equivalent of an oasis in mid-desert, thousands of miles from any significant landmass for trading and replenishing seed. The 300–400 fish species here are fewer than half of what's on display in the great Pacific coral reef diversity triangle of Indonesia, New Guinea, and the Solomon Islands. Yet the pressure of aquarium-trade capture and overfishing by dynamite and cyanide have stressed those places nearly to breaking, and left them bereft of large predators.

"There's no place left in the ocean like the Serengeti that puts it all together," observes Jeremy Jackson.

Yet Kingman Reef, like the Białowieża Puszcza, is a time machine, an intact fragment of what used to surround every green dot in this big blue ocean. Here, the coral team finds a half-dozen unknown species. The invertebrate crew brings back strange mollusks. The microbe team discovers hundreds of new bacteria and viruses, largely because no one has ever before tried to chart the microscopic universe of a coral reef.

In a sweltering cargo hold below decks, microbiologist Forest Rohwer has mini-cloned the lab he runs at San Diego State. Using an oxygen probe just one micron across that's hooked to a microsensor and a laptop, his team has demonstrated exactly how algae that they collected earlier at Palmyra are supplanting living corals. In small glass cubes they built and filled with seawater, they placed bits of coral and seaweed algae separated by a glass membrane so fine that not even viruses can pass through it. Sugars produced by the algae can, however, because they dissolve. When bacteria living on coral feed on this rich extra nutrient, they consume all the available oxygen, and the coral dies.

To verify this finding, the microbiology team dosed some cubes with ampicillin to kill the hyperventilating bacteria, and those corals remained healthy. "In every case," says Rohwer, climbing out of the hold into the considerably cooler afternoon, "stuff dissolving out of the algae kills the coral."

So where is all the weedy algae coming from? "Normally," he explains, lifting his nearly waist-length black hair to catch a breeze on the back of his neck, "coral and algae are in happy equilibrium, with fish grazing on the algae and cropping it. But if water quality around a reef goes down, or if you remove grazing fish from the system, algae get the upper hand."

In a healthy ocean such as at Kingman Reef, there are a million bacteria per milliliter, doing the world's work by controlling the movements of nutrients and carbon through the planet's digestive system. Around the populated Line Islands, however, some samples show 15 times as much bacteria. Taking up oxygen, they choke coral, gaining ground for yet more algae to feed yet more microbial bacteria. It's the slimy cycle that Jeremy Jackson fears, and Forest Rohwer agrees that it could well happen.

"Microbes don't really much care whether we—or anything else—are here or not. We're just a semi-interesting niche for them. In fact, there's been just a very brief period of time when there were anything *but* microbes on the planet. For billions of years, that's all there was. And when the sun starts to expand, we'll go, and it'll only be microbes, for millions or billions of years more."

They will remain, he says, until the sun dries up the last water on Earth, because microbes need it to thrive and reproduce. "Although they can be stored by freeze-drying, and do just fine. Everything we shoot into space has microbes on it, despite people's efforts to not let that happen. Once it's out there, there's no reason why some of this stuff couldn't make it billions of years."

The one thing microbes could never have done was take over the land the way more complex cell structures finally did, building plants and trees that invited more complex life-forms to dwell in them. The only structures microbes create are mats of slime, a regression toward the first life-forms on Earth. To these scientists' palpable relief, here at Kingman that hasn't happened yet. Pods of bottlenose dolphins accompany the dive boats to and from the *White Holly,* leaping to snag plentiful flying fish. Each underwater transect reveals more richness, ranging from gobies, a fish less than a centimeter long, to manta rays the size of Piper Cubs, and hundreds of sharks, snappers, and big jacks.

The reefs themselves, blessedly clean, are lush with table corals, plate corals, lobe corals, brain corals, and flower corals. At times, the walls of coral nearly disappear behind colored clouds of smaller grazing fish. The paradox that this expedition has confirmed is that their sheer abundance is caused by the hordes of hungry predators that devour them. Under such predation pressure, small herbivores reproduce even faster.

"It's like when you mow your lawn," explains Alan Friedlander. "The more you crop it, the more quickly grass grows. If you let it go awhile, the growth rate levels off."

No chance of that happening with all of Kingman's resident sharks. Parrotfish, whose beak-like incisors evolved to gnaw the most tenacious coral-choking algae, even change sex to maintain their sizzling reproduction rates. The healthy reef keeps its system in balance by providing nooks and crevices in which small fish hide long enough to breed before becoming shark food. As a result of the constant conversion of plant and algae nutrients into short-lived little fish, the long-lived apex predators end up accumulating most of the biomass.

The expedition data would later show that fully 85 percent of the live weight at Kingman Reef was accounted for by sharks, snappers, and other carnivorous company. How many PCBs may have migrated up the food chain and now saturate their tissues is fodder for a future study.

Two days before the expedition's scientists depart Kingman, they steer their dive boats to the twin crescent islets heaped atop the northern arm of the boomerang-shaped reef. In the shallows, they find a heartening sight: a spectacular community of spiny black, red, and green sea urchins, robust grazers of algae. A 1998 El Niño temperature fluctuation, ratcheted even higher by global warming, knocked out 90 percent of the sea urchins in the Caribbean. Unusually warm water shocked coral polyps into spitting out friendly algal photosynthesizers that live in tight symbiosis with them, trading just the right balance of sugars for ammonia fertilizer the corals excrete back, and also providing their color. Within a month, more than half the Caribbean reefs had turned to bleached coral skeletons, now coated with slime.

Like corals worldwide, the ones at the Kingman islets' edges also show bleaching scars, but fierce grazing has kept invasive algae at bay, allowing encrusting pink corallines to slowly cement the wounded reef back together. Wading gingerly around all the sea urchin spikes, the researchers climb ashore. Within a few yards, they're on the windward side of the clamshell rubble, where they get a shock.

From one end to the other, each isle is carpeted with crushed plastic bottles, parts of polystyrene floats, nylon shipping ties, Bic lighters, flip-flops in various states of ultraviolet disintegration, plastic bottle caps of assorted sizes, squeeze-tubes of Japanese hand lotion, and a galaxy of multicolored plastic fragments shattered beyond identity.

The only organic detritus is the skeleton of a red-footed booby, chunks of an old wooden outrigger, and six coconuts. The following day, the scientists return after their final dive and fill dozens of garbage bags. They are under no illusions that they have returned Kingman Reef to the pristine state it was in before humans ever found it. Asian currents will bring more plastic; rising temperatures will bleach more corals—possibly all of them, unless coral and its photosynthetic resident algal partner can evolve new symbiotic agreements quickly.

Even the sharks, they now realize, are evidence of human intervention. Only one that they've seen in Kingman all week was a behemoth longer than six feet; the rest are apparently adolescents. Within the past two decades, shark finners must have been here. In Hong Kong, shark fin soup commands up to $100 per bowl. After slicing off their pectoral and dorsal fins, finners throw mutilated sharks, still alive, back into the sea. Rudderless, they sink to the bottom and suffocate. Despite campaigns to ban the delicacy, in less remote waters an estimated 100 million sharks die this way every year. The presence of so many vigorous young ones, at least, gives hope that enough sharks here escaped the blade to revive the population. PCBs or not, they look to be prospering.

"In a year," observes Enric Sala, watching their spotlit frenzy that night from the rail of the *White Holly,* "humans take 100 million sharks, while sharks attack maybe 15 people. This is not a fair fight."

Enric Sala stands on the shore of Palmyra Atoll, waiting for a turbo-prop Gulfstream to land on the airstrip built here the last time the world was officially at war, to take his expedition team back to Honolulu—a three-hour flight. From there, they'll disperse across the globe with their data. When they meet again, it will be electronically, and then in peer-reviewed papers they will coauthor.

Palmyra's soft green lagoon is pure and lucid, its tropical splendor patiently erasing crumbled concrete slabs where thousands of sooty terns now roost. The tallest structure here, a former radar antenna, has rusted in half; within a few more years, it will disappear completely among the coconut palms and almond trees. If all human activity suddenly collapsed along with it, Sala believes that quicker than we'd expect, the reefs of the northern Line Islands could be as complex as they were in the last few thousand years before they were found by men bearing nets and fishhooks. (And rats: probably the onboard, self-reproducing food supply for Polynesian mariners who dared cross this endless ocean with only canoes and courage.)

"Even with global warming, I think reefs would recover within two centuries. It would be patchy. In some places, lots of large predators. Others would be coated with algae. But in time, sea urchins would return. And the fish. And then the corals."

His thick dark eyebrows arch beyond the horizon to picture it. "In 500 years, if a human came back, he'd be completely terrified to jump in the ocean, because there would be so many mouths waiting for him."

Jeremy Jackson, in his sixties, was the elder ecological statesman on this expedition. Most here, like Enric Sala, are in their thirties, and some are younger graduate students. They are of a generation of biologists and zoologists who increasingly append the word *conservation* to their titles. Inevitably, their research involves fellow creatures touched, or simply mauled, by the current worldwide peak predator, their own species. Fifty more years of the same, they know, and coral reefs will look very different. Scientists and realists all, yet their glimpse at how inhabitants of Kingman Reef thrive in the natural balance to which they evolved has only hardened their resolve to restore equilibrium—with humans still around to marvel.

A coconut crab, the world's largest land invertebrate, waddles by. Flashes of pure white amid the almond leaves overhead are the new plumage of fairy tern chicks. Removing his sunglasses, Sala shakes his head.

"I'm so amazed," he says, "by the ability of life to hang on to anything. Given the opportunity, it goes everywhere. A species as creative and arguably intelligent as our own should somehow find a way to achieve a balance. We have a lot to learn, obviously. But I haven't given up on us."

At his feet, thousands of tiny, trembling shells are being resuscitated by hermit crabs. "Even if we don't: if the planet can recover from the Permian, it can recover from the human."

With or without human survivors, the planet's latest extinction will come to an end. Sobering as the current cascading loss of species is, this is not another Permian, or even a rogue asteroid. There is still the sea, beleaguered but boundlessly creative. Even though it will take 100,000 years for it to absorb all the carbon we mined from the Earth and loaded into the air, it will be turning it back into shells, coral, and who knows what else. "On the genome level," notes microbiologist Forest Rohwer, "the difference between coral and us is small. That's strong molecular evidence that we all come from the same place."

Within recent historic times, reefs swarmed with 800-pound groupers, codfish could be dipped from the sea by lowering baskets, and oysters filtered all the water in Chesapeake Bay every three days. The planet's shores teemed with millions of manatees, seals, and walruses. Then, within a pair of centuries, coral reefs were flattened and sea-grass beds were scraped bare, the New Jersey–sized dead zone appeared off the mouth of the Mississippi, and the world's cod collapsed.

Yet despite mechanized overharvesting, satellite fish-trackers, nitrate flooding, and prolonged butchery of sea mammals, the ocean is still bigger than we are. Since prehistoric man had no way to pursue them, it's the one place on Earth besides Africa where big creatures eluded the intercontinental megafaunal extinction. "The great majority of sea species are badly depleted," says Jeremy Jackson, "but they still exist. If people actually went away, most could recover."

Even, he adds, if global warming or ultraviolet radiation bleaches Kingman and Australia's Great Barrier Reef to death, "they're only 7,000 years old. All these reefs were knocked back over and over by the ice ages, and had to form again. If the Earth keeps getting warmer, new reefs will appear farther to the north and south. The world has always changed. It's not a constant place."

Nine-hundred miles northwest of Palmyra, the next visible turquoise-ringed smudge of land rising from the blue Pacific depths is Johnston Atoll. Like Palmyra, it was once a U.S. seaplane base, but in the 1950s it became a Thor missile nuclear test range. Twelve thermonuclear warheads were exploded here; one that failed scattered plutonium debris over the island. Later, after tons of irradiated soil, contaminated coral, and plutonium were "decommissioned" into a landfill, Johnston became a post–Cold War chemical-weapons incineration site.

Until it closed in 2004, sarin nerve gas from Russia and East Germany, along with Agent Orange, PCBs, PAHs, and dioxins from the United States, were burned there. Barely one square mile, Johnston Atoll is a marine Chernobyl and Rocky Mountain Arsenal rolled into one—and like the latter, its latest incarnation is as a U.S. National Wildlife Refuge.

Divers there report seeing angelfish with herringbone chevrons on one side and something resembling a cubist nightmare on the other. Yet, despite this genetic jumble, Johnston Atoll is not a wasteland. The coral seems reasonably healthy, thus far weathering—or perhaps inured to—temperature creep. Even monk seals have joined the tropicbirds and boobies nesting there. At Johnston Atoll, as at Chernobyl, the worst insults we hurl at nature may stagger it, but nowhere as severely as our overindulged lifestyle.

One day, perhaps, we will learn to control our appetites, or our duplication rates. But suppose that before we do, something implausible swoops in to do that for us. In just decades, with no new chlorine and bromine leaking skyward, the ozone layer would replenish and ultraviolet levels subside. Within a few centuries, as most of our excess industrial CO_2 dissipated, the atmosphere and shallows would cool. Heavy metals and toxins would dilute and gradually flush from the system. After PCBs and plastic fibers recycled a few thousand or million times, anything truly intractable would end up buried, to one day be metamorphosed or subsumed into the planet's mantle.

Long before that—in far less time than it took us to run out of codfish and passenger pigeons—every dam on Earth would silt up and spill over. Rivers would again carry nutrients to the sea, where most life would still be, as it was long before we vertebrates first crawled onto these shores.

Eventually, we'd try that again. Our world would start over.

Coda
Our Earth, Our Souls

As the saying goes, we don't get out of this life alive—and neither will the Earth. Around 5 billion years from now, give or take, the sun will expand into a red giant, absorbing all the inner planets back into its fiery womb. At that point, water ice will thaw on Saturn's moon Titan, where the temperature is currently −290°F, and some interesting things may eventually crawl out of its methane lakes. One of them, pawing through organic silt, might come across the Huygens probe that parachuted there from the Cassini space mission in January, 2005, which, during its descent, and for 90 minutes before its batteries died, sent us pictures of streambed-like channels cutting down from orange, pebbled highlands to Titan's sand-dune seas.

Sadly, whatever finds Huygens won't have any clue where it came from, or that we once existed. Bickering among project directors at NASA nixed a plan to include a graphic explanation that Jon Lomberg designed, this time encased in a diamond that would preserve a shred of our story at least 5 billion years—long enough for evolution to provide another audience.

More crucial to us still here on Earth, right now, is whether we humans can make it through what many scientists call this planet's latest great extinction—make it through, and bring the rest of Life with us rather than tear it down. The natural history lessons we read in both the fossil and the living records suggest that we can't go it alone for very long.

Various religions offer us alternative futures, usually elsewhere, although Islam, Judaism, and Christianity mention a messianic reign on Earth lasting, depending on whose version, somewhere from seven to 7,000 years. Since these would apparently follow events that result in severe population reduction of the unrighteous, this might be feasible. (Unless, as all three suggest, the dead would be resurrected, which could trigger both resource and housing crises.)

However, since they disagree about who are the righteous, to believe any one of them requires an act of faith. Science offers no criteria by which to pick survivors other than evolution of the fittest, and into every creed are born similar percentages of stronger and weaker individuals.

As to the fate of the planet and its other residents after we're finally done with it—or it's done with us—religions are dismissive, or worse. The posthuman Earth is either ignored or destroyed, although in Buddhism and Hinduism, it starts again from scratch—as does the whole universe, similar to a repeating Big Bang theory. (Until that happens, the correct answer to whether this world would go on without us, says the Dalai Lama, is: "Who knows?")

In Christianity, the Earth melts, but a new one is born. Since it needs no sun—the eternal light of God and the Lamb having eliminated night—it's clearly a different planet than this one.

"The world exists to serve people, because man is the most honorable of all creatures," says Turkish Sufi master Abdülhamit Çakmut. "There are cycles in life. From the seed comes the tree, from the tree comes the fruit we eat, and we give back as humans. Everything is meant to serve man. If people are gone from this cycle, nature itself will be over."

The Muslim dervish practice he teaches reflects the recognition that everything, from atoms to our galaxy, whirls in cycles, including nature as it continually regenerates—at least until now. Like so many others—Hopis, Hindus, Judeo-Christians, Zoroastrians—he warns of an endtime. (In Judaism, time itself is said to end, but only God knows what that means.) "We see the signs," Çakmut says. "Harmony is broken. The good are outnumbered. There is more injustice, exploitation, corruption, pollution. We are facing it now."

It's a familiar scenario: Good and evil finally spin apart, landing in heaven and hell, respectively, and everything else vanishes. Except, Abdül-hamit Çakmut adds, we can slow this process—the good are those who strive to restore harmony and speed nature's regeneration.

"We take care of our bodies to live a longer life. We should do the same for the world. If we cherish it, make it last as long as possible, we can postpone the judgment day."

Can we? Gaia theorist James Lovelock prophesies that unless things change soon, we'd better stash essential human knowledge at the poles in a medium that doesn't require electricity. Yet Dave Foreman, founder of Earth First!, a cadre of environmental guerrillas who had all but given up on humans deserving a place in the ecosystem, now directs The Rewilding Institute, a think tank based on conservation biology and unapologetic hope.

That hope both includes, and depends on, the consecration of "mega-linkages"—corridors that would span entire continents, where people would be committed to coexisting with wildlife. In North America alone, he sees a minimum of four: they would span the continent's dividing spine, the Atlantic and Pacific coasts, and the Arctic-boreal. In each, top predators and large fauna absent since the Pleistocene would be reinstated, or the closest things possible: African surrogates of America's missing camels, elephants, cheetahs, and lions.

Dangerous? The payoff for humans, Foreman and company believe, is that in a re-equilibrated ecosystem, there's a chance for us to survive. If not, the black hole into which we're shoving the rest of nature will swallow us as well.

It's a plan that keeps Paul Martin, author of the Blitzkrieg extinction theory, in touch with Kenya's David Western, fighting to stop elephants from downing every last drought-stressed fever tree: Send some of those proboscids to America, pleads Martin. Let them again eat Osage oranges, avocados, and other fruit and seeds that evolved to be so big because megafauna could ingest them.

Yet the biggest elephant of all is a figurative one in the planet-sized room that is ever harder to ignore, although we keep trying. Worldwide, every four days human population rises by 1 million. Since we can't really grasp such numbers, they'll wax out of control until they crash, as has happened to every other species that got too big for this box. About the only thing that could change that, short of the species-wide sacrifice of voluntary human extinction, is to prove that intelligence really makes us special after all.

The intelligent solution would require the courage and the wisdom to put our knowledge to the test. It would be poignant and distressing in ways, but not fatal. It would henceforth limit every human female on Earth capable of bearing children to one.

The numbers resulting from such a draconian measure, fairly applied, are tricky to predict with precision: Fewer births, for example, would lower infant mortality, because resources would be devoted to protecting each precious member of the latest generation. Using the United Nations' medium scenario for life expectancy through 2050 as a benchmark, Dr. Sergei Scherbov, who is the research group leader at the Vienna Institute of Demography of the Austrian Academy of Sciences and an analyst for the World Population Program, calculated what would happen to human population if, from now on, all fertile women have only one child (in 2004, the rate was 2.6 births per female; in the medium scenario that would lower to about two children by 2050).

If this somehow began tomorrow, our current 6.5 billion human population would drop by 1 billion by the middle of this century. (If we continue as projected, it will reach 9 billion.) At that point, keeping to one-child-per-human-mother, life on Earth for all species would change dramatically. Because of natural attrition, today's bloated human population bubble would not be reinflated at anything near the former pace. By 2075, we would have reduced our presence by almost by half, down to 3.43 billion, and our impact by much more, because so much of what we do is magnified by chain reactions we set off through the ecosystem.

By 2100, less than a century from now, we would be at 1.6 billion: back to levels last seen in the 19th century, just before quantum advances in energy, medicine, and food production doubled our numbers and then doubled us again. At the time, those discoveries seemed like miracles. Today, like too much of any good thing, we indulge in more only at our peril.

At such far-more-manageable numbers, however, we would have the benefit of all our progress plus the wisdom to keep our presence under control. That wisdom would come partly from losses and extinctions too

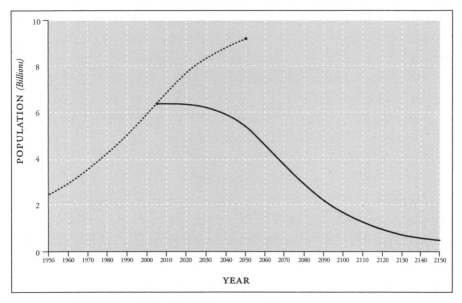

Projections, Population of the World:
----------- Medium UN scenario: fertility declining from 2.6 children per woman in 2004 to slightly over 2 children per woman in 2050. Source: Population Division of the Department of Economic and Social Affairs of the United Nations Secretariat (2005).
----------- Scenario assuming that all fertile women are henceforth limited to 1 child. Source: Dr. Sergei Scherbov, research group leader, Vienna Institute of Demography, Austrian Academy of Sciences.

CHART BY JONATHAN BENNETT

late to reverse, but also from the growing joy of watching the world daily become more wonderful. The evidence wouldn't hide in statistics. It would be outside every human's window, where refreshed air would fill each season with more birdsong.

If we don't, and thus let our numbers increase by half again as projected, could our technology stretch resources once more, as it did for a while back in the 20th century? We've already heard from the robot contingent. Relaxing on the deck of the *White Holly*, watching the sharks roll by, microbiologist Forest Rohwer takes a stab at another theoretical possibility:

"We could try using lasers, or some similar particle-wave beams, to actually build things remotely on other planets, or in other solar systems.

That would be much faster than actually sending something there. Maybe we could code for a human, and build a human in space. The life sciences will probably provide the ability to do that. Whether physics will allow it, I don't know. But it's all just biochemistry, so there's no reason why we couldn't construct it.

"Unless," he allows, "there really is something called a spark of life. But it's going to take something like that, because there's no evidence that we could actually move from here in any reasonable time frame."

If we could do that—find a fertile planet somewhere big enough for all of us, holographically clone our bodies, and upload our minds across light-years—eventually the Earth would do fine without us. With no more herbicides, weeds (otherwise known as biodiversity) would invade our industrial farms and our vast monocultured commercial pine plantations— although, in America, for a while the weeds may mostly be kudzu. It's only been around since 1876, when it was brought from Japan to Philadelphia as a centennial gift to the United States, and eventually something is bound to learn to eat it. In the meantime, without gardeners endlessly trying to uproot the ravenous stuff, long before the vacant houses and skyscrapers of America's southern cities tumble, they may have already vanished under a bright, waxy green, photosynthesizing blanket.

Since the late 19th century, when, beginning with electrons, we got down to manipulating the most fundamental particles of the universe, human life has changed very fast. One measure of how fast is that, barely a century ago—until Marconi's wireless and Edison's phonograph—all the music ever heard on Earth was live. Today, a tiny fraction of 1 percent is. The rest is electronically reproduced or broadcast, along with a trillion words and images each day.

Those radio waves don't die—like light, they travel on. The human brain also emanates electric impulses at very low frequencies: similar to, but far weaker than, the radio waves used to communicate with submarines. Paranormalists, however, insist that our minds are transmitters that, with special effort, can focus like lasers to communicate across great distances, and even make things happen.

That may seem far-fetched, but it's also a definition of prayer.

The emanations from our brains, like radio waves, must also keep going—where? Space is now described as an expanding bubble, but that architecture is still a theory. Along its great mysterious interstellar curvatures, perhaps it's not unreasonable to think that our thought waves might eventually find their way back here.

Or even that one day—long after we're gone, unbearably lonely for the beautiful world from which we so foolishly banished ourselves—we, or our memories, might surf home aboard a cosmic electromagnetic wave to haunt our beloved Earth.

ACKNOWLEDGMENTS

᠁

In July of 2003, tired of watching drought, bark beetles, and fires devour chunks of the Arizona forests I had long called home, I escaped to what I hoped would be more benign weather in upstate New York. The night I arrived at a friend's cabin, so did the first tornado to ever hit the Catskills. The next day, while we were discussing how to remove a seven-foot-long section of spruce sapling that had skewered the eaves like a javelin, I got a message from Josie Glausiusz.

Josie, an editor at *Discover Magazine*, had recently reread an article I'd written for *Harper's* years earlier, describing how when humans fled Chernobyl, nature rushed in to fill our void. Plutonium or not, the ecosystem surrounding the ruined reactor seemed better off without us. "What," she asked me, "would happen if humans disappeared *everywhere?*"

It was a deceptively simple question that, I came to understand, lets us view our Earth's current myriad stresses from the disarming vantage of a fantasy in which we supposedly no longer exist, yet somehow we get to watch what unfolds next. Watch, and maybe learn. The article Josie asked me to write led me to this book-length attempt to address her question more thoroughly, and I am ever grateful to her for posing it.

My agent, Nicholas Ellison, not only sensed a fascinating book in it, but found me the right editor. John Parsley at Thomas Dunne Books/ St. Martin's Press has provided continual reassurance, especially when my research inevitably led to some dark places. I'm indebted to Nick and

John, not only for their professional skill and counsel, but also for always helping me recall why I was writing this.

To produce a book that credibly evokes how our world would continue without people is to enter a paradox that ranks with any Buddhist conundrum: it can't be done without the help of a large supporting cast of humans. Many have appeared in these pages, and I owe them enormously for helping me understand our planet through their eyes, hearts, and expertise. There were many others whose vital contributions are not obvious in the narrative, if only for the sake of economy: had I included everyone's instructive stories, this volume would have been four times heavier than it is.

Among them, I deeply thank Peter Jessop of Integrity Development and Construction in Amherst, Massachusetts, and his design colleagues Anna Novey, Kyle Wilson, and Ben Goodale. Along with Amherst architects Chris Riddle and Laura Fitch, they explained details I'd never dreamed of about the kind of frame structures in which I've lived much of my life. Similarly, a day spent with architect Erin Moore and Arizona State Museum antiquities conservation specialist Chris White, walking through another place I've called home, Tucson, was both illuminating and humbling to realize how much of my own surroundings I'd never fully seen before. In New York City, landscape architects Laura Starr and Stephen Whitehouse, who had just guided the redesign of Battery Park, offered many insights and even more questions that I needed to probe in order to address the fate of buildings, infrastructure, and landscaping without constant human maintenance.

I also thank Brooklyn Botanical Garden's Steve Clemants, who spent some patient hours enlightening me, as did New York Botanical Garden's Dennis Stevenson, Chuck Peters, and herbarium director Barbara Thiers. Across the parkway at the Bronx Zoo, Eric Sanderson and his Mannahatta Project gave me continual inspiration. Charles Seaton of New York Transit arranged my trek through the subways, which Paul Schuber and Peter Briffa ably and affably guided. I spent yet more hours with Cooper Union's civil engineering chairman Jameel Ahmad, and with NYU's poly-talented scientist Tyler Volk and physicist Marty Hoffert. And, courtesy of

Jerry Del Tufo, I now appreciate that a bridge is so much more than simply a means to the other side.

Outer space is as far a stretch from home as we can make; it's my good fortune, however, to have had a genuine rocket scientist for a neighbor. Astrophysicist Jonathan Lunine of the University of Arizona is responsible for much of the thrilling work that has brought us images and understanding of the outer planets. He has the gift of explaining hugely complex cosmic matters in language that not just a college freshman, but even I can understand, and to him I owe the idea of using the *I Love Lucy* show to demonstrate the trajectory of radio signals.

Previous assignments had taken me to some locations that became settings for this book, but several others I had never seen before. In each, I am beholden to people whose knowledge, patience, and generosity added up to an enthralling education.

In Ecuador, I thank cousins Gloria, Bartolo, and Luciano Ushigua: a new generation of Zápara leaders, resurrecting their people.

Seeing Poland's and Belarus's ancient Białowieża Puszcza/Belovezhskaya Pushcha felt like entering holy ground. It is a pilgrimage that I wish every European might make, lest this unparalleled natural heirloom be overwhelmed. My thanks to Andrzej Bobiec, Bogdan Jaroszewicz, and Heorhi Kazulka, not only for showing it to me, but for their exemplary courage and principles.

On the beautiful, sadly divided island of Cyprus, I toured the Green Line courtesy of Wlodek Cibor of the United Nations Peacekeeping Force in Cyprus. Asu Muhtaroglu of the Ministry of Foreign Affairs, Turkish Cyprus, and botanist Mustafa Kemal Merakli showed me Varosha, Karpaz, and much more, as did artist and horticulturist Hikmet Uluçan. In Kyrenia, I thank Kenan Atakol of the CEVKOVA Environmental Protection Trust, Bertil Wedin, Felicity Alcock, and the late Allan Cavinder—and, for invaluable advice and introductions, a longtime former Cyprus resident: American classical guitarist, journalist, and novelist Anthony Weller.

In Turkey, I am deeply obliged for the help and imagination of another novelist, Elif Shafak, who also introduced me to journalists Eyüp Can and David Judson, editors at the Istanbul newspaper *Referans*. Eyüp, in turn, connected me to columnist Metin Münir. All these people gorged me with more wonderful ideas, food, drink, and friendship than I knew a human could hold: learning that it's possible is one of the blessings of field research. In Cappadocia, my excellent guide, Ahmet Sezgin, took me to the Nevşehir Museum to meet archaeologist Murat Ertuğrul Gülyaz, another new friend I intend to keep. Yet another fine journalist, Melis Şenerdem, translated my conversation with Sufi master Abdülhamit Çakmut of the Mevlâna Educational and Cultural Society. After having the honor of witnessing his dervish disciples whirl, I am beholden to him for their reminder that humans are capable of not just Earthly, but ethereal beauty.

David "Jonah" Western's contribution to this book not only came from days of stimulating dialogue—and from the pilot's seat in his Cessna— but also from a generation of colleagues he has inspired to conserve his beloved east equatorial African ecosystems. For their kindness and many good ideas, I thank Samantha Russell and Zippy Wanakuta of the African Conservation Centre; Evans Mgwani of the University of Nairobi; and Dr. Helen Gichohi of the African Wildlife Foundation.

Chicago Tribune correspondent Paul Salopek provided many helpful suggestions about African sites for this book. Long dinner conversations in Nairobi with Kelly West of the World Conservation Union's Regional Office for Eastern Africa and Oscar Sims of Envision Multimedia were critical to connecting African environmental issues to the book's theme. In Kenya, several guides and naturalists showed me places and wildlife that I could never have discovered on my own: David Kimani, Francis Kahuta, Vincent Kiama, Joe Njenga, Joseph Motongu, John Ahalo, Tsavo deputy park warden Kathryn Wambani and education director Lucy Makosi, and, in Maasai Mara, Lemeria Nchoe and Partois ole Santian.

In Tanzania, I thank Joseph Bifa at Olduvai Gorge and Browny Mtaki, who showed me key locales in Serengeti. At Lake Tanganyika, Karen Zwick and Michael Wilson at the Jane Goodall Institute in Kigoma and Gombe were immeasurably knowledgeable and hospitable, and worth traveling days to talk to. Late one evening, NYU doctoral candidate Kate

Detwiler expounded on a theory I'd long puzzled over. I especially thank limnologist Andy Cohen of the University of Arizona for suggestions that led me to all of them, and for sharing his deep experience in the region.

My permission to visit the Korean Demilitarized Zone was kindly and swiftly processed by the U.S. Forces in Korea and the Republic of Korea Army. My preparation owed greatly to Dr. George Archibald of the International Crane Foundation and his colleagues at the DMZ Forum: Hall Healy, Dr. E. O. Wilson of Harvard, and Dr. Kim Ke Chung of Pennsylvania State University. In South Korea, I was hosted with consummate thoroughness by the Korean Federation of Environmental Movement, among the most impressive NGOs I've ever encountered. I fondly thank my traveling companions Ahn Chang-Hee, Kim Kyung-won, Park Jong-Hak, Jin Ik-Tae, and especially Ma Yong-Un, one of the most thoughtful, capable, and committed human beings I have the pleasure to know.

In England, I discovered a true living jewel 30 miles north of the Tower of London: Rothamsted Research. My thanks to Paul Poulton for showing Rothamsted's magnificent archive and its long-running experiments to me, and to Richard Bromilow and Steve McGrath for discussing their work with soil additives and contaminants. Farther south, my understanding of the landscape owes much to traveling through Dartmoor with Tavistock archaeologist Tom Greeves and to a conversation with University of Exeter geographer Chris Caseldine. And, on a beach on Britain's southern coast, Richard Thompson of the University of Plymouth launched me into an investigation of plastics that became one of this book's most lasting—in all senses of the word—metaphors for unintended consequences. My thanks to him and his student Mark Browne, and to plastics experts in the United States he suggested: Tony Andrady of the North Carolina Research Triangle and Capt. John Moore of the Algita Marine Research Foundation.

To visit the petrochemical complex that stretches from Houston to Galveston is both easy and maddeningly difficult. The easy part is that you can't miss it, because in that bend of the Texas Gulf Coast, it is practically omnipresent. The hard part, for whatever proprietary or less-justifiable reasons, is gaining entrance to petroleum and chemical plants. Journalists are regarded much like contaminants—an understandable, but regrettable, defensive reflex. I am grateful to Juan Parras at Texas Southern University for legwork he did on my behalf, and for the openness and candor with which I was eventually received at Texas Petrochemical by environmental-health-safety director Max Jones, and at Valero Refining in Texas City by spokesman Fred Newsome. In the same region, several scientists and ecologists gave me a glimpse of the world before—and possibly after—the human race's potent, but problematic, affair with petroleum derivatives: John Jacob at the Texas Coastal Watershed Program, Brandon Crawford of The Nature Conservancy, Sammy Ray at Texas A&M-Galveston, and, especially, wetlands biologist Andy Sipocz of Texas Parks and Wildlife.

At the Rocky Flats National Wildlife Refuge, I thank Karen Lutz of U.S. Fish and Wildlife; the DOE's Joe Leguerre; and John Rampe, John Corsi, and Bob Nininger of Kaiser-Hill. At the former Rocky Mountain Arsenal, my appreciation goes to refuge manager Dean Rundle and to Matt Kales. Panamanian anthropologist Stanley Heckadon Moreno of the Smithsonian Tropical Research Institute gave me an ecological context for the monumental reality of the Panama Canal that Abdiel Pérez, Modesto Echevers, Johnny Cuevas, and Bill Huff kindly showed me. In Northwest Territories, Arctic guide and pilot "Tundra" Tom Faess flew and hiked me through fabulous parts of the Canadian wilderness, including the diamond-mining region, and the BHP (now BHP-Billiton) Corporation graciously gave me a tour of their Ekati diamond mine, and also a singular thrill: holding a 52-million-year-old chunk of unpetrified redwood in my hands.

As a boy I'd always planned to be a scientist, though I could never figure out what kind, because everything interested me. How could I become an astronomer, if it meant not being a paleontologist? My great fortune as a journalist has been the chance to commune with brilliant scientists from so many disciplines, and in so many fascinating places. Accompanying archaeologist Arthur Demarest to Dos Pilas in Guatemala was among the most memorable trips of my life. Another was visiting Chernobyl with nuclear physicists Andriy Demydenko and Volodya Tykhyy, and landscape architect David Hulse, systems analyst Kit Larsen, and the late, deeply missed environmental educator John Baldwin of the University of Oregon. On an assignment to Antarctica several years ago, courtesy of the National Science Foundation and the *Los Angeles Times Magazine,* optical physicist Ray Smith, biologist Barbara Prezelin, both of the University of California-Santa Barbara, and molecular biologist Deneb Karentz of UC-San Francisco shared their pioneering research on ozone depletion, which remains critical to our understanding today. Over multiple trips to the Amazon, herpetologist Bill Lamar has continually taught me more. Closer to home, I was profoundly moved to see the Harvard Forest with David Foster and old-growth woods in Oregon with U.S. Forest Service geologist Fred Swanson and philosopher and nature writer Kathleen Dean Moore.

I'm still savoring my conversation with the Smithsonian's extinction expert, Doug Erwin. My gratitude for willingness to explain years of scientific sleuthing also goes to research-fisheries biologist Diana Papoulias; ethnobotanist Gary Paul Nabhan; hazardous-materials specialist Enrique Medina; risk assessment engineer Bob Roberts; Stanford "garbologist" William Rathje; paleoornithologist David Steadman, who found the last ground sloths in Caribbean caves; ornithologist Steve Hilty, whose exhaustive bird guides have added weight to both my luggage and my words; and biologist-anthropologist Peter Warshall, who lucidly connects everything. Nuclear-safety engineer David Lochbaum of the Union of Concerned Scientists and nuclear operations and engineering director Alex Marion of the Nuclear Energy Institute were both essential to my understanding of the inner sanctums of nuclear plants. Thanks as well to NEI spokesman Mitch Singer, to Susan Scott of the U.S. Department of Energy's Waste Isolation Pilot Project, and to Arizona Public Service for access to the Palo Verde Nuclear Generating Station. My great admiration, too, to Gregory Benford, the University of California-Irvine physicist and Nebula Award–winning science fiction author who helped me think about time, past and future—no small task.

Paleontologist Richard White has helped Tucson's International Wildlife Museum, which he directs, grow into a research and educational facility— not unlike many renowned museums whose exhibits were originally big-game trophies collected by hunters. I was first taken there by the eminent paleoecologist Paul Martin, who calls it a place of reflection. My very special thanks to Paul Martin for many engrossing hours and enlightening thoughts, and for suggestions springing from his deep familiarity with the canon of scientific literature on extinctions, including many works that challenge his own theories. My final interview on this subject, with

C. Vance Haynes, helped me to put all the competing scholarship into a context that reveals the collective contributions of each.

I can't adequately express my appreciation to Jeremy Jackson and Enric Sala for inviting me to join the Scripps Institution of Oceanography's 2005 expedition to the South Pacific's Line Islands—and for months of conversation and education, before and after. So many scientists on that trip taught me so much that to select only a few to sketch that voyage in this book's final chapter in no way reflects how grateful I am to all. My thanks to marine ecologist and co–principal investigator Stuart Sandin of Scripps; microbiologists Rob Edwards, Olga Pantos, and especially Forest Rohwer of San Diego State; Philippine invertebrate biologist Machel Malay; coral reef specialists David Obura of the Indian Ocean's CORDIO program and Jim Maragos of U.S. Fish and Wildlife; ichthyologists Edward DeMartini of the National Oceanic and Atmospheric Administration and Alan Friedlander of Hawaii's Oceanic Institute; University of California-Santa Barbara marine botanist Jennifer Smith; coral disease expert Liz Dinsdale of Australia's James Cook University; and two Scripps graduate students en route to vital careers: Steve Smriga and Melissa Roth. My education also benefited from the presence on that voyage of Smithsonian diving safety officer Mike Lang, filmmaker Soames Summerhays, and photographer Zafer Kizilkaya. Ecologist Alex Wegman, based on Palmyra, was my helpful source for terrestrial atoll ecology. Last, thank you to Capt. Vincent Backen and his crew of the OR/V *White Holly*, whose skill and hospitality allowed all the science to proceed.

I owe much of my understanding of both methane clathrates and carbon sequestration to Charles Bryer, Hugh Guthrie, and Scott Klara of the National Energy and Technology Laboratory in Morgantown, West Virginia, and to David Hawkins of the National Resources Defense Council. Susan Lapis of SouthWings and Judy Bonds of the Coal River Mountain Watch showed me the ghosts of former mountains in West Virginia, and what it takes to face and fight such devastation. My thanks also to codirector Monica Moore of the Pesticide Action Network of North America for information on the health impacts of agricultural chemistry; to head scientist David Olson at the Colorado School of Mines for his help with metal-alloy longevity; and to planetary scientist Carolyn Porco of the

Cassini Project for thoughts on images literally out of this world. Dr. Thomas Ksiazek, chief of the Special Pathogens Branch at the Centers for Disease Control, and Dr. Jeff Davis, Wisconsin chief medical officer and state epidemiologist for communicable diseases, brought me back to Earth with their impressive dedication to sobering work. Thanks, too, to Michael Mathews of the University of Minnesota and Michael Wilk of Wayne State University for explaining the intricacies of mortuary science, and to Michael Pazar of Wilbert Funeral Services.

Both in discussion and through his always-surprising writing, Oxford's Nick Bostrom challenged my thinking on multiple subjects. I likewise thank Rabbis Michael Grant and Baruch Clein, Rev. Rodney Richards, Todd Strandberg of Rapture Ready, Sufi Abdülhamit Çakmut, Rev. Hyon Gak Sunim, and His Holiness the Dalai Lama for sharing their varied, thought-provoking contemplations of the Earth after us. Each professes one of the world's great religions, but what filled my own soul most was their common humanity—a quality also shared by VHEMT's Les U. Knight, who would bring nature's human experiment to a close, and the Rewilding Institute's Dave Foreman, who would keep us, but in cooperation, not conflict, with the rest of our planet's species. I am particularly beholden to Dr. Wolfgang Lutz of the World Population Program, and his colleague Dr. Sergei Scherbov of the Vienna Institute of Demography at the Austrian Academy of Sciences, for assistance with translating a critical element of that formula into plain numbers—numbers that, quite literally, we could live with. All of us.

My grateful appreciation goes to Jacqueline Sharkey, head of the University of Arizona Journalism Department, and to the university's Center for Latin American Studies, for encouraging me to combine my annual international journalism seminar with my research in Panama. Likewise, my Ecuador trip, where I was greatly assisted by guest producer Nancy Hand, was supported by my bosom partners at Homelands Productions, who are also my constant inspirations: Sandy Tolan, Jon Miller, and Cecilia Vaisman.

Many other friends, relations, and colleagues sustained me at key moments through the research and writing of this book, with contributions ranging from practical to intellectual, moral, and mystical (to say nothing of culinary)—all prompting ideas and bolstering my energy when I most needed them. For advice, critiques, insights, affection, faith, food, and spare bedrooms, thank you Alison Deming, Jeff Jacobson, Marnie Andrews, Drum Hadley, Rebecca West, Mary Caulkins, Karl Kister, Jim Schley, Barry Lopez, Debra Gwartney, Chuck Bowden, Mary Martha Miles, Bill Wing, Terri Windling, Bill Posnick, Pat Lanier, Constanza Vieira, Diana Hadley, Tom Miller, Ted Robbins, Barbara Ferry, Dick Kamp, Jon Hipps, Caroline Corbin, Clark Strand, Perdita Finn, Molly Wheelwright, Marvin Shaver, and Joan Kravetz, and very special thanks to my able research aide, Julie Kentnor. This list also includes some entire families: Nubar Alexanian, Rebecca Koch, and Abby Koch Alexanian; Karen, Benigno, Elias, and Alma Sánchez-Eppler; and Rochelle, Peter, Brian, and Pahoua Hoffman.

I am indebted, too, to the artists whose work graces these pages. Digital magician Markley Boyer brings data to stunning life for the Mannahatta Project. Janusz Korbel has long photographed the splendors of Poland's Białowieża Puszcza for the same impassioned reasons that inspire Vivian Stockman to document the missing mountains of West Virginia. Archaeologist Murat Ertuğrul Gülyaz and biologist Jim Maragos each contributed images evoking their respective subsurface specialities: the underground cities of central Turkey and the Pacific coral reef. *Arizona Republic* photographer Tom Tingle provided an interior glimpse of a realm that few would dare enter, but one to which we are literally connected daily—the core of a nuclear generating reactor.

Peter Yates's image of decaying Varosha, Cyprus, has special poignancy: three decades earlier, he met his wife there. Perhaps symbolically, as he snapped the picture a foreground sprig of wild grass blew in front of his lens, partly obscuring the abandoned hotel's façade; with his assent, it was photo-surgically removed by Ronn Spencer of 'Sole Studios. Ronn and his colleague Blake Hines also expertly processed color photographs into black-and-white renditions for this book.

The reproduction of Annapolis illustrator Phyllis Saroff's resurrected passenger pigeon in flight doesn't fully reflect her delicately colored original, but the gray-scale version she kindly provided has its own haunting power. And I can never thank Carl Buell enough for creating original drawings of a litoptern, a giant sloth, and of our forebear *Australopithecus*, for this book.

Artist Jon Lomberg's contribution here goes far beyond the reproduced silhouette he drew for the interstellar Voyager spacecrafts. Jon's vision exemplifies how art can literally soar beyond our supposed limitations and surprise us with manifestations of spirit that feel linked to eternity. His act of preserving sounds and images that embody such spirit is truly one of humanity's enduring achievements. I am deeply thankful to him, and to Manhattan art conservators Barbara Appelbaum and Paul Himmelstein, for what they bring not only to this book, but to us all.

At MetalPhysic, their Tucson studio and foundry, Tony Bayne and Jay Luker preserve human expression in that most enduring of metal alloys, bronze. I met them through a sculptor who, to my miraculous good fortune, is my wife, Beckie Kravetz. To know that bronze sculptures such as the graceful figures she conjures have a better chance than nearly anything else we humans do of lasting until the end of Earthly time feels utterly fitting and proper to me. Here's a vast understatement: without her, this book simply would not exist.

Here's another: All of us humans have myriad other species to thank. Without them, *we* couldn't exist. It's that simple, and we can't afford to ignore them, any more than I can afford to neglect my precious wife—nor the sweet mother Earth that births and holds us all.

Without us, Earth will abide and endure; without her, however, we could not even be.

—*Alan Weisman*

SELECT BIBLIOGRAPHY

BOOKS

Addiscott, T. M. *Nitrate, Agriculture, and the Environment*. Wallingford, Oxfordshire, U.K.: CABI Publishing, 2005.

Andrady, Anthony, editor. *Plastics and the Environment*. Hoboken: John Wiley & Sons, Inc., 2003.

Audubon, John James. *Ornithological Biography, or an Account of the Habits of the Birds of the United States of America*. Edinburgh: Adam Black, 1831.

Benford, Gregory. *Deep Time*. New York: Avon Books, 1999.

Bobiec, Andrzej. *Preservation of a Natural and Historical Heritage as a Basis for Sustainable Development: A Multidisciplinary Analysis of the Situation in* Białowieża *Primeval Forest, Poland*. Narewka, Poland: Society for Protection of the Białowieża Primeval Forest (TOPB), 2003.

Cantor, Norman. *In the Wake of the Plague: The Black Death, and the World It Made*. New York: Free Press, 2001.

Colborn, Theo, John Peterson Myers, and Dianne Dumanoski. *Our Stolen Future: Are We Threatening Our Own Fertility, Intelligence, and Survival?—A Scientific Detective Story*. New York: Dutton, 1996.

Colinvaux, Paul. *Why Big Fierce Animals Are Rare: An Ecologist's Perspective*. Princeton, N.J.: Princeton University Press, 1978.

Cronon, William. *Changes in the Land: Indians, Colonists, and the Ecology of New England*. New York: Hill and Wang, 1983.

Cronon, William, editor. *Uncommon Ground: Rethinking the Human Place in Nature.* New York: W. W. Norton & Company, 1995.

Crosby, Alfred W. *Ecological Imperialism (Second Edition).* Cambridge: Cambridge University Press, 2004.

Demarest, Arthur. *Ancient Maya: The Rise and Fall of a Rainforest Civilization.* Cambridge: Cambridge University Press, 2004.

Department of Economic and Social Affairs, Population Division, United Nations. *World Population Prospects: The 2004 Revision Highlights.* New York: United Nations, February 2005, vi.

Depleted Uranium Education Project. *Metal of Dishonor, Depleted Uranium: How the Pentagon Radiates Soldiers and Civilians with DU Weapons (Second Edition).* New York: International Action Center, 1997.

Dixon, Douglas. *After Man: A Zoology of the Future.* New York: St. Martin's Press, 1981.

Dreghorn, William. *Famagusta and Salamis: A Guide Book.* Lefkosa, Northern Cyprus: K. Rustem and Bros., 1985.

Dreghorn, William. *A Guide to the Antiquities of Kyrenia.* Nicosia: Halkin Sesi, 1977.

Dyke, George V. *John Lawes of Rothamsted: Pioneer of Science, Farming and Industry.* Harpenden: Hoos, 1993.

Erwin, Douglas. *Extinction: How Life on Earth Nearly Ended 250 Million Years Ago.* Princeton, N.J.: Princeton University Press, 2006.

Evans, W. R., and A. M. Manville II, editors. *Transcripts of Proceedings of the Workshop on Avian Mortality at Communication Towers.* August 11, 1999, Ithaca, N.Y.: Cornell University, 2000, published on the internet at http://www.towerkill.com/ and http://migratorybirds.fws.gov/issues/towers/agenda.html.

Flannery, Tim. *The Eternal Frontier: An Ecological History of North America and Its Peoples.* Melbourne: The Text Publishing Company, 2001.

Flannery, Tim. *The Future Eaters: An Ecological History of the Australasian Lands and People.* Sydney: Reed Books/New Holland, 1994.

Foreman, Dave. *Rewilding North America: A Vision for Conservation in The 21st Century.* Washington D.C.: Island Press, 2004.

Foster, David R. *Thoreau's Country: Journey Through a Transformed Landscape.* Cambridge: Harvard University Press, 1999.

Garrett, Laurie. *The Coming Plague.* New York: Farrar, Straus and Giroux, 1994.

Hall, Eric J. *Radiation and Life.* London: Pergamon, 1984.

Hilty, Steven L., and William L. Brown. *Birds of Colombia.* Princeton N.J.: Princeton University Press, 1986.

Hoffecker, John. *Twenty-Seven Square Miles: Landscape and History at Rocky Mountain Arsenal National Wildlife Refuge.* U.S. Fish and Wildlife Service, 2001.

Jefferson, Thomas. *Notes on the State of Virginia, 1787.* Chapel Hill: University of North Carolina Press, 1982.

Kain, Roger, and William Ravenhill, editors. *Historical Atlas of South-West England.* Exeter: University of Exeter Press, 1999.

Koester, Craig. *Revelation and the End of All Things.* Grand Rapids, Mich.: Wm. B. Eerdmans Publishing Company, 2001.

Kurtén, Björn, and Elaine Anderson. *Pleistocene Mammals of North America.* New York: Columbia University Press, 1980.

Kurzweil, Ray. *The Singularity Is Near: When Humans Transcend Biology.* New York: Viking, 2005.

Langewiesche, William. *American Ground: Unbuilding the World Trade Center.* New York: North Point Press, 2002.

Leakey, Richard, and Roger Lewin. *The Sixth Extinction: Patterns of Life and the Future of Humankind.* New York: Doubleday, 1995.

LeBlanc, Steven A. *Constant Battles.* New York: St. Martin's Press, 2003.

Lehmann, Johannes, et al. *Amazonian Dark Earths: Origin, Properties, Management.* Dordrecht; Boston; London: Kluwer Academic, 2003.

Leslie, John. *The End of the World: The Science and Ethics of Human Extinction.* London: Routledge, 1996.

Lovelock, James. *The Ages of Gaia: A Biography of Our Living Earth.* New York: W. W. Norton & Company, 1988.

Lovelock, James. *Gaia: A New Look at Life on Earth.* Oxford: Oxford University Press, 1979.

Lovelock, James. *The Revenge of Gaia.* London: Allen Lane/Penguin Books, 2006.

Lunine, Jonathan I. *Earth: Evolution of a Habitable World.* Cambridge: Cambridge University Press, 1999.

Mann, Charles C. *1491: New Revelations of the Americas Before Columbus*. New York: Alfred A. Knopf, 2005.

Marcó del Pont Lalli, Raúl, editor. *Electrocución de Aves en Líneas Eléctricas de México: Hacia un Diagnóstico y Perspectivas de Solución*. México, D.F.: INE-Semarnat, 2002.

Martin, Paul. *The Last 10,000 Years: A Fossil Pollen Record of the American Southwest*. Tucson, Ariz.: The University of Arizona Press, 1963.

Martin, Paul. *Twilight of the Mammoths: Ice Age Extinctions and the Rewilding of America*. Berkeley, Calif.: University of California Press, 2005.

Martin, Paul, and H. E. Wright, editors. *Pleistocene Extinctions: The Search for a Cause* New Haven, Conn.: Yale University Press, 1967.

McCullough, David. *Path Between the Seas: The Creation of the Panama Canal 1870–1914*. New York: Simon & Schuster, 1977.

McGrath, S. P., and P. J. Loveland. *The Soil Geochemical Atlas of England and Wales*. London: Blackie Academic and Professional, 1992.

McKibben, Bill. *The End of Nature, 10th Anniversary Edition*. New York: Doubleday/Anchor Books, 1999.

Moorehead, Alan. *The Fatal Impact: The Invasion of the South Pacific 1767–1840*. New York: Harper & Row, 1967.

Moulton, Daniel, and John Jacob. *Texas Coastal Wetlands Guidebook*. Texas Parks & Wildlife, no date.

Muller, Charles. *The Diamond Sutra*. Toyo Gakuen University, Copyright 2004, http://www.hm.tyg.jp/~acmuller/bud-canon/diamond_sutra.html.

Mwagore, Dali, editor. *Land Use in Kenya: The Case for a National Land Use Policy*. Nakuru, Kenya: Kenya Land Alliance, no date.

Mycio, Mary. *Wormwood Forest: A Natural History of Chernobyl*. Washington D.C.: Joseph Henry Press, 2005.

Outwater, Alice. *Water: A Natural History*. New York: Basic Books 1996.

Ponting, Clive. *A Green History of the World*. London: Sinclair-Stevenson, 1991.

Potts, Richard. *Humanity's Descent: The Consequences of Ecological Instability*. New York, William Morrow & Co., 1996.

Rackham, Oliver. *Ancient Woodland: Its History, Vegetation and Uses in England*. London: E. Arnold, 1980.

Rackham, Oliver. *The Illustrated History of the Countryside.* London: Weidenfeld & Nicolson Ltd., 1994.

Rackham, Oliver. *Trees and Woodland in the British Landscape.* London: Dent, 1990.

Rathje, William, and Cullen Murphy. *Rubbish! The Archeology of Garbage.* Tucson, Ariz.: University of Arizona Press, 2001.

Rees, Martin. *Our Final Hour.* New York: Basic Books, 2003.

Rothamsted Experimental Station. *Rothamsted: Guide to the Classical Field Experiments.* Harpenden, Hertsfordshire, U.K.: AFRC Institute of Arable Crops Research, 1991.

Safina, Carl. *Eye of the Albatross.* New York: Henry Holt and Company, 2002.

Safina, Carl. *Song for the Blue Ocean.* New York: Henry Holt and Company, 1998.

Sagan, Carl, F. D. Drake, Ann Druyan, Timothy Ferris, Jon Lomberg, and Linda Salzman Sagan. *Murmurs of Earth: The Voyager Interstellar Record.* New York: Random House, 1978.

Schama, Simon. *Landscape and Memory.* New York: Alfred A. Knopf, 1995.

Simmons, Alan. *Faunal Extinction in an Island Society: Pygmy Hippopotamus Hunters of Cyprus.* New York: Kluwer Academic/Plenum Publishers, 1999.

Steadman, David, and Jim Mead, editors. *Late Quaternary Environments and Deep History: A Tribute to Paul Martin.* Hot Springs, S.Dak.: The Mammoth Site of Hot Springs, South Dakota, Inc., 1995.

Stewart, George R. *Earth Abides.* New York, Houghton Mifflin Company, 1949.

Strum, Shirley C. *Almost Human: A Journey into the World of Baboons.* New York: Random House, 1987.

The Texas State Historical Association. *The Handbook of Texas Online.* Austin, Tex.: University of Texas Libraries and the Center for Studies in Texas History, 2005, http://www.tsha.utexas.edu/handbook/online/index.html.

Thomas, Jr., William L. *Man's Role in Changing the Face of the Earth.* Chicago: University of Chicago Press, 1956.

Thorson, Robert M. *Stone by Stone: The Magnificent History in New England's Stone Walls.* New York: Walker & Company, 2002.

Todar, Kenneth. *Online Textbook of Bacteriology.* Madison, Wisc.: University of Wisconsin, Department of Bacteriology, 2006, http://textbookofbacteriology.net.

Turner, Raymond, H. Awala Ochung', and Jeanne Turner. *Kenya's Changing Landscape*. Tucson, Ariz.: University of Arizona Press, 1998.

Wabnitz, Colette, et al. *From Ocean to Aquarium*. Cambridge, U.K.: UNEP World Conservation Monitoring Centre, 2003.

Ward, Peter, and Donald Brownlee. *The Life and Death of Planet Earth*. New York: Henry Holt and Company LLC, 2002.

Ward, Peter, and Alexis Rockman. *Future Evolution*. New York: Times Books, 2001.

Weiner, Jonathan. *The Beak of the Finch: A Story of Evolution in Our Time*. New York: Alfred A. Knopf, 1994.

Western, David. *In the Dust of Kilimanjaro*. Washington, D.C.: Island Press, 1997.

Wilson, Edward. O. *The Diversity of Life (1999 Edition)*. New York: W. W. Norton & Company, 1999.

Wilson, Edward. O. *The Future of Life*. New York: Alfred A. Knopf, 2002.

Wrangham, Richard, and Dale Peterson. *Demonic Males: Apes and the Origins of Human Violence*. New York: Panther/Houghton Mifflin Company, 1996.

Yurttaş, Şükruü. *Cappadocia*. Ankara: Rekmay Ltd., no date.

Zimmerman, Dale, Donald Turner, and David Pearson. *Birds of Kenya and Northern Tanzania,* Princeton, N.J.: Princeton University Press, 1999.

ARTICLES

Advocacy Project. "The Zapara: Rejecting Extinction." *Amazon Oil,* vol. 16, no. 8, March 21, 2002.

Allardice, Corbin, and Edward R. Trapnell. "The First Pile." Oak Ridge, Tenn.: United States Atomic Energy Commission, Technical Information Service, 1955.

Alpert, Peter, David Western, Barry R. Noon, Brett G. Dickson, Andrzej Bobiec, Peter Landres, and George Nickas. "Managing the Wild: Should Stewards Be Pilots?" *Frontiers in Ecology and the Environment,* vol. 2, no. 9, 2004, 494–99.

Andrady, Anthony L. "Plastics and Their Impacts in the Marine Environment." *Proceedings of the International Marine Debris Conference on Derelict Fishing Gear and the Ocean Environment,* August 6–11, 2000 Hawai'i Convention Center, Honolulu, Hawai'i.

Avery, Michael L. "Review of Avian Mortality Due to Collisions with Man-made Structures." *Bird Control Seminars Proceedings,* University of Nebraska, Lincoln, 1979, 3–11.

Avery, Michael, P. F. Springer, and J. F. Cassel. "The Effects of a Tall Tower on Nocturnal Bird Migration—a Portable Ceilometer Study." *Auk,* vol. 93, 1976, 281–91.

Ayhan, Arda. "Geological and Morphological Investigations of the Underground Cities of Cappadocia Using GIS." Master's thesis, Department of Geological Engineering, Graduate School of Natural and Applied Sciences, Middle East Technical University, 2004.

Baker, Allan J., et al. "Reconstructing the Tempo and Mode of Evolution in an Extinct Clade of Birds with Ancient DNA: The Giant Moas of New Zealand." *Proceedings of the National Academy of Sciences,* vol. 102, no. 23, June 7, 2005, 8257–62.

Baker, R. J., and R. K. Chesser. "The Chornobyl Nuclear Disaster and Subsequent Creation of a Wildlife Preserve." *Environmental Toxicology and Chemistry,* vol. 19, 2000, 1231–32.

Barlow, Connie, and Tyler Volk. "Open Living Systems in a Closed Biosphere: A New Paradox for the Gaia Debate." *BioSystems,* vol. 23, 1990, 371–84.

Barnes, David K. A. "Remote Islands Reveal Rapid Rise of Southern Hemisphere, Sea Debris." *The Scientific World Journal,* vol. 5, 2005, 915–21.

Beason, R. C. "The Bird Brain: Magnetic Cues, Visual Cues, and Radio Frequency (RF) Effects." *Proceedings of Conference on Avian Mortality at Communication Towers,* August 11, 1999, Cornell University, Ithaca, N.Y.

Beason, R. C. "Through a Bird's Eye—Exploring Avian Sensory Perception." USDA/Wildlife Services/National Wildlife Research Center, Ohio Field Station, Sandusky, Ohio, http://www.aphis.usda.gov/ws/nwrc/is/03pubs/beason031.pdf.

Bjarnason, Dan. "Silver Bullet: Depleted Uranium." Producer Marijka Hurkol, Canadian Broadcasting Corporation, January 8, 2001.

Blake, L., and K. W. T. Goulding. "Effects of Atmospheric Deposition, Soil pH and Acidification on Heavy Metal Contents in Soils and Vegetation of Semi-Natural Ecosystems at Rothamsted Experimental Station, UK." *Plant and Soil,* vol. 240, 2002, 235–51.

Bobiec, Andrzej. "Living Stands and Dead Wood in the Białowieża Forest: Suggestions for Restoration Management." *Forest Ecology and Management,* vol. 165, 2002, 121–36.

Bobiec, Andrzej., H. van der Burgt, K. Meijer, and C. Zuyderduyn. "Rich Deciduous Forests in Białowieża as a Dynamic Mosaic of Developmental Phases: Premises for Nature Conservation and Restoration Management." *Forest Ecology and Management,* vol. 130, 2000, 159–75.

"Bomb Facts: How Nuclear Weapons Are Made." *Wisconsin Project on Nuclear Arms Control,* November 2001, http://www.wisconsinproject.org/pubs/articles/2001/bomb%20facts.htm.

Bostrom, Nick. "Are You Living in a Computer Simulation?" *Philosophical Quarterly,* vol. 53, no. 211, 2003, 243–55.

Bostrom, Nick. "Existential Risks: Analyzing Human Extinction Scenarios and Related Hazards." *Journal of Evolution and Technology,* vol. 9, March 2002.

Bostrom, Nick. "A History of Transhumanist Thought." *Journal of Evolution and Technology,* vol. 14, no. 1, 2005, 1–25.

Bostrom, Nick. "When Machines Outsmart Humans." *Futures,* vol. 35, no. 7, 2003, 759–64.

Bromilow, Richard H., et al. "The Effect on Soil Fertility of Repeated Applications of Pesticides over 20 Years." *Pesticide Science,* vol. 48, 1996, 63–72.

Butterfield, B. J., W. E. Rogers, and E. Siemann. "Growth of Chinese Tallow Tree (*Sapium sebiferum*) and Four Native Trees Along a Water Gradient." *Texas Journal of Science* [*Big Thicket Science Conference Special Issue*], 2004.

Canine, Craig. "How to Clean Coal." *OnEarth,* Natural Resources Defense Council, fall, 2005.

Cappiello, Dina. "New BP Leak in Texas City Is Third Incident This Year." *Houston Chronicle,* August 11, 2005.

Cappiello, Dina. "Unit at Refinery Has Troubled Past." *Houston Chronicle,* August 11, 2005.

Carlson, Elof Axel. "Commentary: International Symposium on Herbicides in the Vietnam War: An Appraisal." *BioScience,* vol. 30, no. 8, September 1983, 507–12.

Chesser, R. K., et al. "Concentrations and Dose Rate Estimates of [134, 137]Cesium and [90]Strontium in Small Mammals at Chornobyl, Ukraine." *Environmental Toxicology and Chemistry,* vol.19, 1999, 305–12.

Choi, Yul. "An Action Plan for Achieving an Eco-Peace Community on the Korean Peninsula." In Peninsula: *DMZ Ecosystem Conservation.* The DMZ Forum, 2002, 137–42.

Clark, Ezra, and Julian Newman. "Push to the Finishing Line: Why the Montreal Protocol Needs to Accelerate the Phaseout of CFC Production for Basic Domestic Needs." *EIA Briefing 61-1,* Environmental Investigation Agency, July 2003.

Cobb, Kim, et al. "El Niño/Southern Oscillation and Tropical Pacific Climate During the Last Millennium." *Nature,* vol. 424, July 17, 2003, 271–76.

Cohen, Andrew S., et al. "Paleolimnological Investigations of Anthropogenic Environmental Change in Lake Tanganyika." *Journal of Paleolimnology,* vol. 34, 2005, 1–18.

Cole, W. Matson, Brenda E. Rodgers, Ronald K. Chesser, and Robert J. Baker. "Genetic Diversity of *Clethrionomys glareolus* Populations from Highly Contaminated Sites in the Chornobyl Region, Ukraine." *Environmental Toxicology and Chemistry,* vol. 19, no. 8, 2000, 2130–35.

Coleman, J. S., and S. A. Temple. "On the Prowl." *Wisconsin Natural Resources Magazine,* December 1996.

Coleman, J. S., S. A. Temple, and S. R. Craven. "Cats and Wildlife: A Conservation Dilemma." *1997 USFWS and University of Wisconsin Extension Report,* Madison, Wisc, 1997.

"Complaints by Workers Mar Bloom in Flower Farms." *The Nation* (Nairobi), August 24, 2005.

"Contact-Handled Transuranic Waste Acceptance Criteria for the Waste Isolation Pilot Plant Revision 4.0." WIPP/DOE—02-3122, December 29, 2005.

"Contaminants Released to Surface Water from Rocky Flats." Technical Topic Papers, Rocky Flats Historical Public Exposures Studies, Colorado Department of Public Health and Environment, no date.

"Continued Production of CFCs in Europe." Environmental Investigation Agency, September 20, 2005, http://www.eia-international.org/cgi/news/news.cgi?t=template&a=270.

de Bruijn, Onno, Heorhi Kazulka, and Czesław Okolow, editors. "The Bialowieza Forest in the Third Millennium." *Proceedings of the Cross-border Conference held in Kamenyuki (Belarus) and Bialowieza (Poland),* June 27–29, 2000.

DeMartini, Edward. "Habitat and Endemism of Recruits to Shallow Reef Fish Populations: Selection Criteria for No-take MPAs in the NWHI Coral Reef Ecosystem Reserve." *Bulletin of Marine Science,* vol. 74, no. 1, 185–205.

DeMartini, Edward, Alan Friedlander, and Stephani Holzwarth. "Size at Sex Change in Protogynous Labroids, Prey Body Size Distributions, and Apex Predator

Densities at NW Hawaiian Atolls." *Marine Ecology Progress Series,* vol. 297, 2005, 259–71.

de Waal, Frans B. M. "Bonobo Sex and Society." *Scientific American,* March 1995, 82–88.

"Depleted Uranium." World Health Organization Fact Sheet No. 257, Revised January 2003.

Diamond, Jared. "Blitzkrieg Against the Moas." *Science,* vol. 287, no. 5461, March 24, 2000, 2170–71.

Diamond, Steve. "A Brief History of Johnston Atoll." *15th Airlift Wing History Office Web,* Hickam AFB, Hawaii, http://www2.hickam.af.mil/ho/past/JA/JA_history_home.html.

Donlan, Josh, et al. "Re-wilding North America." *Nature,* vol. 436, August 18, 2005, 913–14.

Doyle, Alister. "Arctic 'Noah's Ark' Vault to Protect World's Seeds." *Reuters,* May 30, 2006.

Ellegren, Hans, et al. "Fitness Loss and Germline Mutations in Barn Swallows Breeding in Chernobyl." *Nature,* vol 389, October 9, 1997, 593–96.

Erickson, Wallace P., Gregory D. Johnson, and David P. Young, Jr. "A Summary and Comparison of Bird Mortality from Anthropogenic Causes with an Emphasis on Collisions." *USDA Forest Service General Technical Reports,* PSW-GTR-191, 2005, 1029–42.

Erwin, Douglas. "Impact at the Permo-Triassic Boundary: A Critical Evaluation." Rubey Colloquium Paper, *Astrobiology,* vol. 3, no. 1, 2003, 67–74.

Erwin, Douglas. "Lessons from the Past: Biotic Recoveries from Mass Extinctions." *Proceedings of the National Academy of Sciences,* vol. 98, no. 10, May 8, 2001, 5399–5403.

Evans, Thayer. "Fire Still Smoldering at BP Unit near Alvin; an Investigation to Determine the Cause Must Wait Until Blaze Is Out." *Houston Chronicle,* August 12, 2005.

Fiedel, Stuart, and Gary Haynes. "A Premature Burial: Comments on Grayson and Meltzer's 'Requiem for Overkill.'" *Journal of Archaeological Science,* vol. 31, no. 1, January 2004, 121–31.

Fleming, Andrew. "Dartmoor Reaves." *Devon Archaeology: Dartmoor Issue,* vol. 3, 1985 (reprinted 1991), 1–6.

Foster, David. R. "Land-Use History (1730–1990) and Vegetation Dynamics in Central New England, USA." *Journal of Ecology,* vol. 80, no. 4, December 1992, 753–71.

Foster, David R., Glenn Motzkin, and Benjamin Slater. "Land-Use History as Long-term Broad-Scale Disturbance: Regional Forest Dynamics in Central New England." *Ecosystems,* vol. 1, 1998, 96–119.

Friedlander, Alan, and Edward DeMartini. "Contrasts in Density, Size, and Biomass of Reef Fishes Between the Northwestern and the Main Hawaiian Islands: The Effects of Fishing down Apex Predators." *Marine Ecology Progress Series,* vol. 230, 2002, 253–64.

Galik, K., B. M. Senut, D. Pickford, J. Treil Gommery, A. J. Kuperavage, and R. B. Eckhardt. "External and Internal Morphology of the BAR 1002'00 *Orrorin tugenensis* Femur." *Science,* vol. 305, September 3, 2004, 1450–53.

Gamache, Gerald L., et al. "Longitudinal Neurocognitive Assessments of Ukranians Exposed to Ionizing Radiation After the Chernobyl Nuclear Accident." *Archives of Clinical Neuropsychology,* vol. 20, 2005, 81–93.

Gao, F., et al. "Origin of HIV-1 in the Chimpanzee *Pan troglodytes troglodytes.*" *Nature* vol. 397, February 4, 1999, 436–41.

Gochfeld, Michael. "Dioxin in Vietnam—the Ongoing Saga of Exposure." *Journal of Occupational Medicine,* vol. 43, no. 5, May 1, 2001, 433–34.

Gopnik, Adam. "A Walk on the High Line." *The New Yorker,* May 21, 2001, 44–49.

Graham-Rowe, Duncan. "Illegal CFCs Imperil the Ozone Layer." *New Scientist,* December 17, 2005, 16.

Grayson, Donald K., and David J. Meltzer. "Clovis Hunting and Large Mammal Extinction: A Critical Review of the Evidence." *Journal of World Prehistory,* vol. 16, no. 4, December 2002, 313–59.

Greeves, Tom. "The Dartmoor Tin Industry—Some Aspects of Its Field Remains." *Devon Archaeology: Dartmoor Issue,* vol. 3, 1985 (reprinted in 1991), 31–40.

Grunwald, Michael. "Monsanto Hid Decades of Pollution: PCBs Drenched Ala. Town, But No One Was Ever Told." *Washington Post,* January 1, 2002, online clarification, 1/5/02; clarification corrected 1/11/02, http://www.washingtonpost.com/ac2/wp-dyn?pagename=article&node=&contentId=A46648–2001Dec31.

Gülyaz, Murat Ertuğrul. "Subterranean Worlds." In *Cappadocia.* Istanbul, Ayhan þahenk Foundation, 1998, 512–25.

Gushee, David E. "CFC Phaseout: Future Problem for Air Conditioning Equipment?" *Congressional Research Service,* Report 93-382 S, April 1, 1993.

Habib, Daniel, et al. "Synthetic Fibers as Indicators of Municipal Sewage Sludge, Sludge Products, and Sewage Treatment Plant Effluents." *Water, Air, and Soil Pollution,* vol. 103, no. 1, April 1, 1998, 1–8.

"Halocarbons and Minor Gases." Chapter 5 in *Emissions of Greenhouse Gases in the United States 1987–1992,* Washington, D.C.: Energy Information Administration Office of Energy Markets and End Use, U.S. Department of Energy, DOE/ EIA-0573, October, 1994.

Harmer, Ralph, et al. "Vegetation Changes During 100 Years of Development of Two Secondary Woodlands on Abandoned Arable Land." *Biological Conservation,* vol. 101, 2001, 291–304.

Hawkins, David G. "Passing Gas: Policy implications of leakage from geologic carbon storage sites." In J. Gale and J. Kaya, editors, *Proceedings of the 6th International Conference on Greenhouse Gas Control Technologies.* Kyoto, Japan, October 2002, Amsterdam: Elsevier, 2003.

Hawkins, David. "Stick it Where??—Public Attitudes Toward Carbon Storage." *Proceedings from the First National Conference on Carbon Sequestration,* DOE/National Energy Technology Laboratory, May 2001.

Hayden, Thomas. "Trashing the Oceans." *U.S. News & World Report,* vol. 133, no. 17, November 4, 2002, 58.

Haynes, C. Vance. "The Rancholabrean Termination: Sudden Extinction in the San Pedro Valley, Arizona, 11,000 B.C." In Juliet E. Morrow and Cristóbal Gnecco, editors, *Paleoindian Archaeology: A Hemispheric Perspective.* Gainesville: University Press of Florida, 2006.

Haynes, Gary. "Under Iron Mountain: Corbis Stores 'Very Important Photographs' at Zero Degrees Fahrenheit." *News Photographer,* January 2005.

Herscher, Ellen. "Archaeology in Cyprus." *American Journal of Archaeology,* vol. 99, no. 2, April 1995, 257–94.

Holdaway, R. N., and C. Jacomb. "Rapid Extinction of the Moas (Aves: Dinornithiformes): Model, Test, and Implications." *Science,* vol. 287, no. 5461, March 24, 2000, 2250–54.

Hotz, Robert Lee. "An Eden Above the City." *Los Angeles Times,* May 15, 2004.

Howden, Daniel. "Varosha Doomed to Rot Away in a Lonely Mediterranean Paradise." *The Independent,* April 26, 2004.

Ichikawa, Mitsuo. "The Forest World as a Circulation System: The Impacts of Mbuti Habitation and Subsistence Activities on the Forest Environment." *African Study Monographs,* suppl.26, March 2001, 157–68.

Jackson, Jeremy B. C. "Reefs Since Columbus." *Coral Reefs* 6 (suppl.), 1997, S23–S32.

Jackson, Jeremy B. C. "What Was Natural in the Coastal Oceans?" *Proceedings of the National Academy of Sciences,* May 8, 2001, vol. 98, no. 10, 5411–18.

Jackson, Jeremy B. C., et al. "Historical Overfishing and the Recent Collapse of Coastal Ecosystems." *Science,* vol. 293, July 27, 2001, 629–38.

Jackson, Jeremy B. C., and Kenneth G. Johnson. "Life in the Last Few Million Years." *The Paleontological Society,* 2000, 221–35.

Jackson, Jeremy B. C., and Enric Sala. "Unnatural Oceans." *Scientia Marina 65* (supp. 2), 2001, 273–81.

Jewett, Thomas O. "Thomas Jefferson, Paleontologist." *The Early America Review,* vol. 3, no. 2, fall 2000.

Jin, Y. G., et al. "Pattern of Marine Mass Extinction Near the Permian-Triassic Boundary in South China." *Science,* vol. 289, July 21, 2000.

Joy, Bill. "Why the Future Doesn't Need Us." *Wired,* vol. 8, no. 4, April, 2000.

Kaiser-Hill Company, L.L.C. "Final Draft: Landfill Monitoring and Maintenance Plan and Post Closure Plan, Rocky Flats Environmental Technology Site Present Landfill." January 2006, http://192.149.55.183/NewRelease/PLFMMPlandraft final23Jan061.pdf.

Kassam, Aneesa, and Ali Balla Bashuna. "The Predicament of the Waata, Former Hunter-gatherers of East and Northeast Africa: Etic and Emic Perspectives." Paper presented at the Ninth International Conference on Hunters and Gatherers, Edinburgh, Scotland, September 9–13, 2002.

Katz, Miriam E., et al. "Uncorking the Bottle: What triggered the Paleocene/Eocene thermal maximum methane release?" *Paleoceanography,* vol. 16, no. 6, December 2001, 549–62.

Katzev, S. W. "The Kyrenia Shipwreck: Clue to an Ancient Crime." *The Athenian,* March 1982, 26–28.

Katzev, S. W., and M. L. Katzev. "Last Harbor for the Oldest Ship." *National Geographic,* November 1974, 618–25.

Kazulka, Heorhi. "Belovezhskaya Pushcha: They Go On Logging It Out, On and On and On. . . ." *Narodnaia Volia,* vol. 2, no.1565, January 4, 2003.

Keating, Barbara. "Insular Geology of the Line Islands." In B. H. Keating and B. Bolton, editors, *Geology and Offshore Mineral Resources of the Central Pacific Basin*. Earth Science Monograph Series, Springer Verlag, New York 1992, 77–99.

Keddie, Grant. "Human History: The Atlatl Weapon." Royal BC Museum, Victoria, British Columbia, Canada, N.D., http://www.royalbcmuseum.bc.ca/hhistory/atlatl-1.pdf.

Kerr, Richard A. "At Last, Methane Lakes on Saturn's Icy Moon Titan—But No Seas." Science, vol. 313, August 4, 2006, 398.

Kiehl, Jeffrey, and Christine Shields. "Climate Simulation of the Latest Permian: Implications for Mass Extinction." *Geology*, September 2005, vol. 33, no. 9, 757–60.

Kim, Ke Chung. "Preserving Biodiversity in Korea's Demilitarized Zone." *Science*, vol. 278, no. 5336, October 10, 1997, 242–43.

Kim, Ke Chung. "Preserving the DMZ Ecosystem: The Nexus of Pan-Korean Nature Conservation." In Peninsula: *DMZ Ecosystem Conservation*. The DMZ Forum, 2002, 171–91.

Kim, Ke Chung, and Edward O. Wilson. "The Land That War Protected." *The New York Times Op-Ed*, Tuesday, December 10, 2002, A 31.

Kim, Kew-gon. "Ecosystem Conservation and Sustainable Use in the DMZ and CCA." In Peninsula: *DMZ Ecosystem Conservation*. The DMZ Forum, 2002, 214–50.

Klem, Jr., Daniel. "Bird-Window Collisions." *Wilson Bulletin*, vol.101, no.4, 1989, 606–20.

Klem, Jr., Daniel. "Collisions Between Birds and Windows: Mortality and Prevention." *Journal of Field Ornithology*, vol. 61, no. l, 1990, 120–28.

Klem, Jr., Daniel. "Glass: A Deadly Conservation Issue for Birds." *Bird Observer*, vol. 34, no. 2, 2006, 73–81.

Koppes, Clayton R. "Agent Orange and the Official History of Vietnam." *Reviews in American History*, vol. 13, no. 1, March 1985, 131–35.

Kurzweil, Ray. "Our Bodies, Our Technologies." *Cambridge Forum Lecture*, May 4, 2005, http://www.kurzweilai.net/meme/frame.html?main=/articles/art0649.html.

Kusimba, Chapurukha M., and Sibel B. Kusimba. "Hinterlands and cities: Archaeological investigations of economy and trade in Tsavo, southeastern Kenya." *Nyame Akuma*, no. 54, December 2000, 13–24.

Lenzi Grillini, Carlo R. "Structural analysis of the Chambura Gorge forest (Queen Elizabeth National Park, Uganda)." *African Journal of Ecology*, vol. 38, 2000, 295–302.

Levy, Sharon, "Navigating With A Built-In Compass." *National Wildlife,* vol. 37, no. 6, October–November 1999.

Little, Charles E. "America's Trees Are Dying." *Earth Island Journal,* fall 1995.

Long, Chun-lin, and Jieuru Wang. "Studies of Traditional Tea-Gardens of Jinuo Nationality, China." In S. K. Jain, editor, *Ethnobiology in Human Welfare.* New Delhi, Deep Publications, 1996, pp. 339–44.

Lorenz, R. D., S. Wall, J. Radebaugh, G. Boubin, E. Reffet, M. Janssen, E. Stofan, R. Lopes, R. Kirk, C. Elachi, J. Lunine, K. Mitchell, F. Paganelli, L. Soderblom, C. Wood, L. Wye, H. Zebker, Y. Anderson, S. Ostro, M. Allison, R. Boehmer, P. Callahan, P. Encrenaz, G. G. Ori, G. Francescetti, Y. Gim, G. Hamilton, S. Hensley, W. Johnson, K. Kelleher, D. Muhleman, G. Picardi, F. Posa, L. Roth, R. Seu, S. Shaffer, B. Stiles, S. Vetrella, E. Flamini, and R. West. "The Sand Seas of Titan: Cassini RADAR Observations of Longitudinal Dunes." *Science,* vol. 312, May 5, 2006, 724–27.

Lozano, Juan A. "Recent Accidents at BP Plants Raise Safety Concerns." *Houston Chronicle, Associated Press,* August 11, 2005.

"M919 Cartridge 25mm, Armor Piercing, Fin Stabilized, Discarding Sabot, with Tracer (APFSDS-T)." *Military Analysis Network,* Federation of American Scientists, 1998, http://www.fas.org/man/dod-101/sys/land/m919.htm.

Markowitz, Michael. "The Sewer System." *Gotham City Gazette,* October 20, 2003.

Martin, Paul S., and D. W. Steadman. "Prehistoric Extinctions on Islands and Continents." In R. D. E. MacPhee, editor, *Extinctions in Near Time: Causes, Contexts and Consequences.* New York: Kluwer/Plenum Press, 1999, 17–55.

Martin, Paul S., and Christine R. Szuter. "War Zones and Game Sinks in Lewis and Clark's West." *Conservation Biology,* vol. 13, no. 1, February 1999, 36–45.

Mato, Y., et al. "Plastic Resin Pellets as a Transport Medium for Toxic Chemicals in the Marine Environment." *Environmental Science and Technology,* vol. 35, 2001, 318–24.

Mayell, Hillary. "Fossil Pushes Upright Walking Back 2 Million Years, Study Says." *National Geographic News,* September 2, 2004.

Mayell, Hillary. "Ocean Litter Gives Alien Species an Easy Ride." *National Geographic News,* April 29, 2002.

McGrath, S. P. "Long-term Studies of Metal Transfers Following Applications of Sewage Sludge." In P. J. Coughtrey, M. H. Martin, and M. H. Unsworth, *Pollutant Transport and Fate in Ecosystems.* Special Publication No. 6 of the British Ecological Society, Oxford: Blackwell Scientific, 1987, 301–17.

McRae, Michael. "Survival Test for Kenya's Wildlife." *Science,* vol. 280, no. 5363, April 28, 1998.

Michel, Thomas. "100 Years of Groundwater Use and Subsidence in the Upper Texas Gulf Coast." Groundwater Reports, Texas Water Development Board, 2005, 139–48.

Milling, T. J. "Leak of gas sends dozens to hospital." *Houston Chronicle,* May 9, 1994.

Mineau, Pierre, and Mélanie Whiteside. "Lethal Risk to Birds from Insecticide Use in the United States—a Spatial and Temporal Analysis." *Environmental Toxicology and Chemistry,* vol. 25, no. 5, 2006, 1214–22.

Ministry of the Environment of Japan. "Report: ODS Recovery and Disposal Workshop in Asia and the Pacific Region." Siem Reap, Cambodia, November 6, 2004.

Ministry of the Environment of Japan. "Revised Report of the Study on ODS [Ozone-Depleting Substances] Disposal Options in Article 5 Countries." May 2006.

Møller, Anders Pape, et al. "Condition, Reproduction and Survival of Barn Swallows from Chernobyl." *Journal of Animal Ecology* vol.74, 2005, 1102–11.

Møller, Anders Pape, and Timothy A. Mousseau. "Biological Consequences of Chernobyl: 20 Years On." *Trends in Ecology and Evolution,* vol. 21, no. 4, April 2006, 200–207.

"Monte Verde Under Fire." *Online Features,* Archaeological Institute of America, October 18, 1999, http://www.archaeology.org/online/features/clovis/.

Moore, Charles. "Trashed: Across the Pacific Ocean, Plastics, Plastics, Everywhere." *Natural History Magazine,* vol. 112, no. 9, November 2003.

Moore, Charles. "A Comparison of Plastic and Plankton in the North Pacific Central Gyre." *Marine Pollution Bulletin,* vol. 42, no. 12, 2001, 1297–1300.

Moore, Charles., et al. "A Brief Analysis of Organic Pollutants Sorbed to Pre- and Post-Production Plastic Particles from the Los Angeles and San Gabriel River Watersheds." *Proceedings of the Plastic Debris, Rivers to Sea Conference, Redondo Beach, CA, Sept. 2005,* http://conference.plasticdebris.org/proceedings.html.

Moore, Charles., et al. "A Comparison of Neustonic Plastic and Zooplankton Abundance in Southern California's Coastal Waters." *Marine Pollution Bulletin,* vol. 44, 2002, 1035–38.

Moore, Charles., et al. "Density of Plastic Particles found in zooplankton trawls from Coastal Waters of California to the North Pacific Central Gyre." *Proceedings of the*

Plastic Debris, Rivers to Sea Conference, Redondo Beach, CA, Sept. 2005, http://conference.plasticdebris.org/proceedings.html.

Moore, Charles., et al. "Working Our Way Upstream: A Snapshot of Land-Based Contributions of Plastic and Other Trash to Coastal Waters and Beaches of Southern California." *Proceedings of the Plastic Debris, Rivers to Sea Conference, Redondo Beach, CA, Sept. 2005,* http://conference.plasticdebris.org/proceedings.html.

Moran, Kevin. "15th Body Pulled from Refinery Rubble." *Houston Chronicle,* March 24, 2005.

Moran, Kevin, and Bill Dawson. "Painful encounter: Leak of Toxic Chemicals Sends Texas City Residents Scurrying." *Houston Chronicle,* April 2, 1998.

Moss, C. J. "The Demography of an African Elephant (*Loxodonta africana*) Population in Amboseli, Kenya." *Journal of Zoology,* 2001, vol. 255, 145–56.

Mullen, Lisa. "Piecing Together a Permian Impact." *Astrobiology Magazine,* May 13, 2004, http://www.astrobio.net/news/modules.php?op=modload&name=News&file=article&sid=969.

Myers, Norman, and Andrew Knoll. "The Biotic Crisis and the Future of Evolution." *Proceedings of the National Academy of Sciences,* May 8, 2001, vol. 98, no. 10, 5389–92.

Norris, Robert S., and William M. Arkin. "Global nuclear stockpiles, 1945-2000." *The Bulletin of the Atomic Scientists,* vol. 56, no. 02, March–April 2000, 79.

Norris, Robert S., and Hans Kristensen "Nuclear Weapons Data: NRDC Nuclear Notebook." *The Bulletin of the Atomic Scientists,* 2006, http://www.thebulletin.org/nuclear_weapons_data/.

Norton, M. R., H. B. Shah, M. E. Stone, L. E. Johnson, and R. Driscoll. "Overview—Defense Waste Processing Facility Operating Experience." Westinghouse Savannah River Company, WSRC-MS-2002-00145, Contract No. DE-AC09-96SR18500, U.S. Department of Energy.

Ochego, Hesbon. "Application of Remote Sensing in Deforestation Monitoring: A Case Study of the Aberdares (Kenya)." Presented at the 2nd FIG Regional Conference, Marrakech, Morocco, December 2–5, 2003.

Oh, Jung-Soo. "Biodiversity and Conservation Stratgies in the DMZ and CCA." In Peninsula: *DMZ Ecosystem Conservation.* The DMZ Forum, 2002, 192–213.

Olivier, Susanne, et al. "Plutonium from Global Fallout Recorded in an Ice Core from the Belukha Glacier, Siberian Altai." *Environmental Science and Technology,* vol. 38, no. 24, 2004, 6507–12.

O'Reilly, Catherine M., Simone R. Alin, Pierre-Denis Plisnier, Andrew S. Cohen, and Brent A. McKee. "Climate Change Decreases Aquatic Ecosystem Productivity of Lake Tanganyika, Africa." *Nature,* vol. 424, August 14, 2003, 766–68.

OSPAR Commission. "Convention for the Protection of the Marine Environment of the North-East Atlantic." Paris, September 21–22, 1992.

Overpeck, Jonathan, et al. "Paleoclimatic Evidence for Future Ice-Sheet Instability and Rapid Sea-Level Rise." *Science,* March 24, 2006, vol. 311, 1747–50.

Owen, James. "Oceans Awash with Microscopic Plastic, Scientists Say." *National Geographic News,* May 6, 2004.

Pandolfi, J. M., R. H. Bradbury, E. Sala, T. P. Hughes, K. A. Bjorndal, R. G. Cooke, D. Macardle, L. McClenahan, M. J. H. Newman, G. Paredes, R. R. Warner, and J. B. C. Jackson. "Global Trajectories of the Long-term Decline of Coral Reef Ecosystems." *Science,* vol. 301, August 15, 2005, 955–58.

Pandolfi, J. M., J. B. C. Jackson, N. Baron, R. H. Bradbury, H. M. Guzman, T. P. Hughes, C. V. Kappel, F. Micheli, J. C. Ogden, H. P. Possingham, and E. Sala. "Are U.S. Coastal Reefs on a Slippery Slope to Slime?" *Science,* vol. 307, March 18, 2005, 1725–26, supporting online material: www.sciencemag.org/cgi/content/full/307/5716/1725/DC1.

Peters, Charles M., et al. "Oligarchic Forests of Economic Utilization and Conservation of Tropical Resource." *Conservation Biology,* vol. 3, no. 4, December 1989, 341–49.

Piller, Charles. "An Alert Unlike Any Other." *Los Angeles Times,* May 3, 2006.

Pinsker, Lisa M. "Applying Geology at the World Trade Center Site." *Geotimes,* vol. 46, no. 11, November 2001.

Potts, Richard. "Complexity and Adaptability in Human Evolution." Manuscript submitted to the American Academy of Arts and Sciences, in association with the July 2001 conference "Development of the Human Species and Its Adaptation to the Environment," http://www.uchicago.edu/aff/mwc-amacad/biocomplexity/conference_papers/PottsComplexity.pdf.

Potts, Richard, et al. "Field Dispatches, The Olorgesailie Prehistoric Site: A Joint Venture of the Smithsonian Institution and the National Museums of Kenya, June 22–August 18, 2004." http://www.mnh.si.edu/anthro/humanorigins/aop/olorg2004/dispatch/start.htm.

Potts, Richard, Anna K. Behrensmeyer, Alan Deino, Peter Ditchfield, and Jennifer Clark. "Small Mid-Pleistocene Hominin Associated with East African Acheulean Technology." *Science,* vol. 305, no. 2, July 2004, 75–78.

Poulton, P. R., et al. "Accumulation of Carbon and Nitrogen by Old Arable Land Reverting to Woodland." *Global Change Biology*, vol. 9, 2003, 942–55.

Quammen, David. "Spirit of the Wild." *National Geographic*, vol. 208, September 2005, 122–43.

Quammen, David. "The Weeds Shall Inherit the Earth." *The Independent*, November 22, 1998, 30–39.

"Radioactive Waste." U.S. Nuclear Regulatory Commission, http://www.nrc.gov/ waste.html, March 1, 2006.

"Raising the Quality: Treatment and Disposal of Sewage Sludge." Department for Environment, Food and Rural Affairs (U.K.), September 23, 1998, 13.

Reaney, Patricia. "Cultivated Land Disappears in AIDs-ravaged Africa."*Reuters*, September 8, 2005.

"Report of Workshop of Experts from Parties to the Montreal Protocol to Develop Specific Areas and a Conceptual Framework of Cooperation to Address Illegal Trade in Ozone-Depleting Substances." Montreal, April 3, 2005, United Nations Environment Programme.

"Reprocessing and Spent Nuclear Fuel Management at the Savannah River Site." Institute for Energy and Environmental Research, Takoma Park, Maryland, February 1999.

Richardson, David, and Remy Petit. "Pines as Invasive Aliens: Outlook on Transgenic Pine Plantations in the Southern Hemisphere." In Claire G. Williams, editor, *Landscapes, Genomics and Transgenic Conifer Forests*. New York: Springer Press, 2005.

Richkus, Kenneth D., et al. "Migratory bird harvest information, 2004 Preliminary Estimates." U.S. Fish and Wildlife Service. U.S. Department of the Interior, Washington, D.C., 2005.

Roach, John. "Are Plastic Grocery Bags Sacking the Environment?" *National Geographic News*, September 2, 2003.

Rodda, Gordon H., Thomas H. Fritts, and David Chiszar. "The Disappearance of Guam's Wildlife." *Bioscience*, vol. 47, no. 9, October 1997, 565–75.

Rodgers, Brenda E., Jeffrey K. Wickliffe, Carleton J. Phillips, Ronald K. Chesser, and Robert J. Baker. "Experimental Exposure of Naive Bank Voles (*Clethrionomys glareolus*) to the Chornobyl, Ukraine, Environment: a Test of Radioresistance." *Environmental Toxicology and Chemistry*, vol. 20, no. 9, 2001, 1936–41.

Rogoff, David. "The Steinway Tunnels." *Electric Railroads*, no. 29, April 1960.

Rubin, Charles T. "Artificial Intelligence and Human Nature." *The New Atlantis*, no. 1, spring 2003, 88–100.

Ruddiman, W. F. "Ice-Driven CO_2 Feedback on Ice Volume." *Climate of the Past*, vol. 2, 2006, 43–55.

Sala, Enric, and George Sugihara. "Food web theory provides guidelines for marine conservation." In Andrea Belgrano, et al., *Aquatic Food Webs: An ecosystem approach*. Oxford: Oxford University Press, 2005, 170–83.

Sanderson, E. W., M. Jaiteh, M. E. Levy, et al. "The Human Footprint and the Last of the Wild." *BioScience*, vol. 52, 2002, 891–904.

Sapolsky, Robert M. "A Natural History of Peace." *Foreign Affairs*, January–February, 2006.

Savidge, Julie A. "Extinction of an Island Forest Avifauna by an Introduced Snake." *Ecology*, vol. 68, no. 3, 1987, 660–68.

Schecter, Arnold, Le Cao Dai, Olaf Päpke, Joelle Prange, John D. Constable, Muneaki Matsuda, Vu Duc Thao, and Amanda L. Piskac. "Recent Dioxin Contamination From Agent Orange in Residents of a Southern Vietnam City." *Journal of Occupational Medicine*, vol. 43, no.5, May 1, 2001, 435–43.

Scherbov, Dr. Sergei, Research Group Leader, Vienna Institute of Demography. "World, Total Population: Assumption Is That from Now on All Women Have One Child." Unpublished, personal communication with author, June 15, 2006.

Scientific American Discovering Archaeology Special Report: Monte Verde Revisited. November–December 1999:

> Fiedel, Stuart J. "Artifact Provenience at Monte Verde: Confusion and Contradictions," 1–12.
> Dillehay, Tom, et al. "Reply to Fiedel, Part I," 12–14.
> Collins, Michael. "Reply to Fiedel, Part II," 14–15.
> West, Frederick H. "The Inscrutable Monte Verde," 16–17.
> Haynes, Vance. "Monte Verde and the Pre-Clovis Situation in America," 17–19.
> Anderson, David G. "Monte Verde and the Way American Archaeology Does Business," 19–20.
> Adovasio, J. M. "Paradigm-Death and Gunfights," 20.
> Bonnichsen, Robson. "A Little Kinder?" 20–21.
> Tankersley, Ken B. "The Truth Is Out There!" 21–22.

Sheldrick, Daphne. "The Elephant Debate." The David Sheldrick Wildlife Trust, 2006, http://www.sheldrickwildlifetrust.org/html/debate.html.

Smith, Thierry, Kenneth D. Rose, and Philip D. Gingerich. "Rapid Asia-Europe-North America Geographic Dispersal of Earliest Eocene Primate *Teilhardina* During the Paleocene-Eocene Thermal Maximum." *Proceedings of the National Academy of Sciences USA*, vol. 103, no. 30, July 25, 2006, 11223–27.

Spinney, Laura. "Return to Paradise." *New Scientist*, vol. 151, no. 2039, July 20, 1996, 26.

Steadman, David. "Prehistoric Extinctions of Pacific Island Birds: Biodiversity Meets Zooarchaeology." *Science*, vol. 267, 1123–31.

Steadman, David, et al. "Asynchronous extinction of late Quaternary sloths on continents and islands." *Proceedings of the National Academy of Sciences USA*, vol. 102, no. 33, August 16, 2005, 11763–68.

Steadman, David, G. K. Pregill, and S. L. Olson. "Fossil vertebrates from Antigua, Lesser Antilles: Evidence for late Holocene human-caused extinctions in the West Indies." *Proceedings of the National Academy of Sciences USA*, vol. 81, 1984, 4448–51.

Steadman, David, and Anne Stokes. "Changing Exploitation of Terrestrial Vertebrates During the Past 3000 Years on Tobago, West Indies." *Human Ecology*, vol. 30, no. 3, September 2002, 339–67.

Stengel, Marc K. "The Diffusionists Have Landed." *The Atlantic Monthly*, January 2000, vol. 285, no. 1, 35–48.

Sterling, Bruce. "One Nation, Invisible." *Wired*, Issue 7.08, August 1999.

Stevens, William K. "New Suspect in Ancient Extinctions of the Pleistocene Megafauna: Disease." *New York Times*, April 29, 1997.

Stewart, Jr., C. Neal, et al. "Transgene Introgression from Genetically Modified Crops to Their Wild Relatives." *Nature*, vol. 4, Oct. 2003, 806–17.

Sublette, Carey. "The Nuclear Weapon Archive: A Guide to Nuclear Weapons." May 2006, http://nuclearweaponarchive.org/.

Takada, Hideshige. "Pellet Watch: Global Monitoring of Persistent Organic Pollutants (POPs) Using Beached Plastic Resin Pellets." *Proceedings of the Plastic Debris, Rivers to Sea Conference, Redondo Beach, CA Sept. 2005*, http://conference.plasticdebris.org/proceedings.html.

Tamaro, George J. "World Trade Center 'Bathtub': From Genesis to Armageddon." *The Bridge (National Academy of Engineering)*, vol. 32, no. 3, Spring 2002, 11–17.

"Technical Factsheet on: Lead." In *National Primary Drinking Water Regulations*. U.S. Environmental Protection Agency, February 28, 2006, http://www.epa.gov/OGWDW/dwh/t-ioc/lead.html.

Tegmark, Max, and Nick Bostrum. "How Unlikely Is a Doomsday Catastrophe?" *Nature*, December 8, 2005, vol. 438, 754.

Thompson, Clive. "Derailed." *New York Magazine*, February 28, 2005.

Thompson, Daniel Q., Ronald L. Stuckey, and Edith B. Thompson. "Spread, Impact, and Control of Purple Loosestrife (*Lythrum salicaria*) in North American Wetlands." U.S. Fish and Wildlife Service. Jamestown, N.Dak.: Northern Prairie Wildlife Research Center, online, June 4, 1999, http://www.npwrc.usgs.gov/resource/plants/loosstrf/loosstrf.htm.

Thompson, Richard C., et al. "Lost at Sea: Where Is All the Plastic?" *Science*, vol. 304, May 7, 2004, 838.

Thorson, Robert M. "Stone Walls Disappearing." *Connecticut Woodlands*, Winter 2005.

"ToxFAQs for Polycyclic Aromatic Hydrocarbons (PAHs)." Agency for Toxic Substances and Disease Registry, 1996, http://www.atsdr.cdc.gov/tfacts69.html.

U.S. Army Environmental Policy Institute (USAEPI), "Health and Environmental Consequences of Depleted Uranium Use by the U.S. Army." Summary Report to Congress, June 1994.

van der Linden, Bart, Harm Smeenge, and Frank Verhart. *Sustainable Forest Degeneration in Bialowieza*, 2004, http://www.franknature.nl.

Vartanyan, S. L., et al. "Radiocarbon Dating Evidence for Mammoths on Wrangel Island, Arctic Ocean, Until 2000 BC." *Radiocarbon*, vol. 37, no. 1, 1995, 1–6.

Vitello, Paul. "Rusty Railroad on Its Way to Pristine Park." *New York Times*, June 15, 2005.

Volk, Tyler. "Sensitivity of Climate and Atmospheric CO_2 to Deep-Ocean and Shallow-Ocean Carbonate Burial." *Nature*, vol. 337, 1989, 637–40.

Wagner, Thomas. "Humans in England May Go Back 700,000 Years." *Associated Press*, December 14, 2005.

Weinstock, J., E. Willerslev, A. Sher, W. Tong, S. Y. Ho, et al. "Evolution, Systematics, and Phylogeography of Pleistocene Horses in the New World: A Molecular Perspective." *Public Library of Science: Biology*, vol. 3, no. 8, August 2005, e241.

Weisman, Alan. "Diamonds in the Wild." *Condé Nast Traveler*, December 2001, 104+.

Weisman, Alan. "Earth Without People." *Discover Magazine*, vol. 26, no. 02, February 2005, 60–65.

Weisman, Alan. "Journey Through a Doomed Land." *Harper's,* vol. 289, no. 1731, August 1994, 45–53.

Weisman, Alan. "Naked Planet." *Los Angeles Times Magazine,* April 5, 1992, 16+.

Weisman, Alan. "The Real Indiana Jones." *Los Angeles Times Magazine,* October 14, 1990, 12+.

Wenning, Richard J., et al. "Importance of Implementation and Residual Risk Analyses in Sediment Remediation." *Integrated Environmental Assessment and Management,* vol. 2, no. 1, 59–65.

Wesołowski, Tomasz. "Virtual Conservation: How the European Union Is Turning a Blind Eye to Its Vanishing Primeval Forest." *Conservation Biology,* vol. 19, no. 5, October 2005, 1349–58.

Western, David. "Human-modified Ecosystems and Future Evolution."*Proceedings of the National Academy of Sciences,* May 8, 2001, vol. 98, no. 10, 5458–65.

Western, David, and Manzolillo Nightingale. "Environmental change and the vulnerability of pastoralists to drought: The Maasai in Amboseli, Kenya." *Africa Environment Outlook Case Studies,* United Nations Environment Programme, Nairobi, 2003.

Western, David, and Manzolillo Nightingale. "Keeping the East African Rangelands Open and Productive." *Conservation and People,* vol. 1, no. 1, October 2005.

Westling, Arthur, et al. "Long-term Consequences of the Vietnam War: Ecosystems." *Report to the Environmental Conference on Cambodia, Laos and Vietnam,* September 15, 2002.

Willis, Edwin O. "Populations and Local Extinctions of Birds on Barro Colorado Island, Panama." *Ecological Monographs,* vol. 44, no. 2, spring 1974, 153–69.

"WIPP Remote-Handled Transuranic Waste Study." DOE/CAO 95–1095, U.S. Department of Energy, Carlsbad Area Office, Carlsbad, N. Mex., October 1995.

Yamaguchi, Eiichiro. "Waste Tire Recycling." Master's thesis, Theoretical and Applied Mechanics, University of Illinois at Urbana, Champaign, October 2000, http://www.p2pays.org/ref/11/10504/.

INDEX

❧

Page numbers in **boldface** refer to illustrations.